Training Load, Well-Being, and Readiness: Reducing Injury Risk and Improving Sports Performance

Training Load, Well-Being, and Readiness: Reducing Injury Risk and Improving Sports Performance

Editor

Filipe Manuel Clemente

MDPI • Basel • Beijing • Wuhan • Barcelona • Belgrade • Manchester • Tokyo • Cluj • Tianjin

Editor
Filipe Manuel Clemente
Escola Superior de Desporto e Lazer
Instituto Politécnico de Viana do Castelo
Melgaço
Portugal

Editorial Office
MDPI
St. Alban-Anlage 66
4052 Basel, Switzerland

This is a reprint of articles from the Special Issue published online in the open access journal *Medicina* (ISSN 1648-9144) (available at: www.mdpi.com/journal/medicina/special_issues/Training_sport).

For citation purposes, cite each article independently as indicated on the article page online and as indicated below:

LastName, A.A.; LastName, B.B.; LastName, C.C. Article Title. *Journal Name* **Year**, *Volume Number*, Page Range.

ISBN 978-3-0365-3867-9 (Hbk)
ISBN 978-3-0365-3868-6 (PDF)

Contents

About the Editor

Filipe Manuel Clemente

Filipe Manuel Batista Clemente has been a university professor since the 2012/2013 academic year and is currently an assistant professor at Escola Superior de Desporto e Lazer de Melgaço (IPVC, Portugal). Filipe holds a Ph.D. in Sports Sciences –Sports Training from the University of Coimbra; his dissertation entitled "Towards a new approach to match analysis: understanding football players'synchronization using tactical metrics"involved observation and match analysis in soccer

As scientific merit, Filipe has had 264 articles published and/or accepted by journals indexed with an impact factor (JCR), as well as over 105 scientific articles that have been peer-reviewed indexed in other indexes. In addition to scientific publications in journals and congresses, he is also the author of six international books and seven national books in the areas of sports training and football. He has also edited various special editions subordinate to sports training in football in journals with an impact factor and/or indexed in SCImago. Additionally, he is a frequent reviewer for impact factor journals in quartiles 1 and 2 of the JCR.

Although he started producing research in 2011, he was included in the restricted list of the world's most-cited researchers in the world (where only eight other Portuguese researchers in sports sciences appear), which was published in the journal Plos Biology in 2020. In 2021, the list was updated, with Filipe Manuel Clemente being again included in the top 2% of the world researchers, in which was positioned in the second place in six Portugueses included in the area of sports sciences. Filipe M. Clemente's SCOPUS h-index is 24 (with a total of 2605 citations), and his Google h-index is 35 (5305 citations). In a list promoted by independent website Expert Escape, he was ranked 40th of 14,875 researchers of football (soccer) in 2020 and in 19th of 15949 in 2021.

Preface to "Training Load, Well-Being, and Readiness: Reducing Injury Risk and Improving Sports Performance"

Monitoring an athlete's performance is part of a sports training puzzle that aims to decrease injury risk and improving the ultimate performance. Therefore, applying a consistent monitoring system involves individually tracking the internal and external load of athletes, analyzing the impact of accumulated load and peripherical factors on well-being and assessing the athlete's readiness. These main topics allow coaches and sports scientists to make decisions about the more adequate training approaches to implement in the athletes. An individual training process depends on well-implemented monitoring systems that provide important information about the thresholds of increasing or decreasing the load, managing the athlete, and reducing the injury risk associated with erroneous training plans. Despite too much information provided by monitoring systems, it still needs more and better research that helps to bring more accurate and precise information that helps to understand critical points or identify the most appropriate approaches based on the interaction between factors.Therefore, this Special Issue "Training load, well-being, and readiness: reducing injury risk and improving sports performance"provided an opportunity to publish original articles, systematic reviews, and/or meta-analyses conducted in the topics of (i) training load monitoring and interactions with fitness, injury risk, and sports performance; (ii) well-being monitoring and interactions with training load, lifestyle, nutrition, and psychological factors; and (iii) readiness assessment and interactions with fatigue, training load, and well-being.

Filipe Manuel Clemente
Editor

Article

Effects of Knee Injury Length on Jump Inside Kick Performances of Wushu Player

Jun-Youl Cha [1], Ha-Sung Lee [2], Sihwa Park [3,*] and Yong-Seok Jee [3,*]

1 Division of Sports & Guard, Howon University, Gunsan 54058, Korea; roksfcha@hanmail.net
2 Department of Education (Physical Education Major), Graduate School of Education of Hanseo University, Seosan 31962, Korea; kamui123@hanmail.net
3 Research Institute of Sports and Industry Science, Hanseo University, Seosan 31962, Korea
* Correspondence: slim@korea.ac.kr (S.P.); jeeys4314@gmail.com (Y.-S.J.); Tel.: +82-41-660-1088 (S.P.); +82-41-660-1028 (Y.-S.J.)

Abstract: *Background and Objectives*: When performing the jump inside kick in Wushu, it is important to understand the rotation technique while in mid-air. This is because the score varies according to the mid-air rotation, and when landing after the mid-air rotation, it causes considerable injury to the knee. This study aimed to compare the differences in kinematic and kinetic variables between experienced and less experienced knee injuries in the Wushu players who perform 360°, 540°, and 720° jump inside kicks in self-taolu. *Materials and Methods*: The participants' mean (SD) age was 26.12 (2.84) years old. All of them had suffered knee injuries and were all recovering and returning to training. The group was classified into a group with less than 20 months of injury experience (LESS IG, n = 6) and a group with more than 20 months of injury experience (MORE IG, n = 6). For kinematic measurements, jump inside kicks at three rotations were assessed by using high-speed cameras. For kinetic measurements, the contraction time and maximal displacement of tensiomyography were assessed in the vastus lateralis, vastus medialis, rectus femoris, biceps femoris, gastrocnemius lateralis, gastrocnemius medialis, and tibialis anterior. The peak torque, work per repetition, fatigue index, and total work of isokinetic moments were assessed using knee extension/flexion, ankle inversion/eversion, and ankle plantarflexion/dorsiflexion tests. *Results*: Although there was no difference at the low difficulty level (360°), there were significant differences at the higher difficulty levels (540° and 720°) between the LESS IG and the MORE IG. For distance and time, the LESS IG had a shorter jump distance, but a faster rotation time compared to those in the MORE IG. Due to the characteristics of the jump inside kick's rotation to the left, the static and dynamic muscle contractility properties were mainly found to be higher in the left lower extremity than in the right lower extremity, and higher in the LESS IG than in the MORE IG. In addition, this study observed that the ankle plantarflexor in the LESS IG was significantly higher than that in the MORE IG. *Conclusion*: To become a world-class self-taolu athlete while avoiding knee injuries, it is necessary to develop the static and dynamic myofunctions of the lower extremities required for jumping. Moreover, it is considered desirable to train by focusing on the vertical height and the amount of rotation during jumping.

Keywords: injury experience; jump inside kick; static muscle contractility; dynamic muscle contractility; ankle plantarflexor

Citation: Cha, J.-Y.; Lee, H.-S.; Park, S.; Jee, Y.-S. Effects of Knee Injury Length on Jump Inside Kick Performances of Wushu Player. *Medicina* **2021**, *57*, 1166. https://doi.org/10.3390/medicina57111166

Academic Editor: Filipe Manuel Clemente

Received: 21 September 2021
Accepted: 26 October 2021
Published: 27 October 2021

1. Introduction

Wushu is an elite sport that is also practiced as a Chinese martial art. The number of Wushu practitioners in Korea is slowly increasing and many athletes have won medals since Wushu was established as an official sport in the Beijing Asian Games in 1990. In 2003, the rules, content, and judging method were changed and new competitions, known as routine-taolu and self-taolu, were introduced [1,2]. In 2005, the Macao East Asian Games began to apply self-taolu, which allows individual athletes to showcase their own skills and abilities in a creative way. The competition of self-taolu should represent the most effective movements in a short amount of time for each event on a specialized 8 m × 14 m carpet [3]. The scoring method involves ten judges who are divided into Groups A, B, and C. Group A can score five points (quality of operation), Group B can score three points (level of performance), and Group C can score two points (1.4 points for technical difficulty, and 0.6 points for connection difficulty). The level of difficulty, divided into A, B, and C levels, is critical in determining the competition performance of the jump kick motions. In order to obtain a high score in the self-taolu competition, it is necessary to be able to perform techniques of high skill levels. For doing this, many spins must be performed in the air. To date, the spinning skills in self-taolu consist of 360°, 540°, and 720° rotations in the air [4,5].

Jump actions in the self-taolu are skills that include running movements just before jumping, kicking in mid-air, landing posture after performing the jump kick, and quick connection with the following technique immediately after landing. Most injuries occur during landing after jumping, and the knee is reported to be the most common among lower extremity joints [6]. Meanwhile, mid-air moves can only be performed well if an athlete has learned scientific training methods and acquired skills according to other difficult moves. For the kinematic aspects of mid-air motions, the velocity change of the body center in performing the 540° rotation was analyzed. Aerial motion can only show altitude and the internal rotation of the left and right lower extremity joints around the body vertical axis at the highest apex [3,7]. It is desirable to effectively utilize ground reactions during rolling, and the right foot should not distribute forces in the left and right directions to lower the center of gravity and increase the height of the jump [5].

The duration of being in mid-air, the rotation of the upper body before jumping, and the action of the outside or inside kicks during the jump are important in determining the level of difficulty of the jump inside kick [8]. A study reported that it is beneficial to increase acceleration by applying the ground reaction force in the hop motion for the jump outside kick [7]. Increasing the force on the footplate by lengthening the hop time also makes it possible to efficiently ascend into the air with greater force [9,10]. It is important to understand the techniques, such as running, hopping, ground reaction force, jumping, and spinning time, as well as the kinematic posture of high-ranked athletes to achieve the required physiological abilities. However, there is a lack of evidence in the literature regarding the analysis of the Wushu jump inside kick for self-taolu. Moreover, there is a lack of research on exercise performance in Wushu players, especially the myofunction of the lower extremities, of athletes who returned to the field after a knee injury.

Therefore, this study investigated the performance of 360°, 540°, and 720° jump inside kicks in Wushu self-taolu for athletes who had injuries related to the aerial rotation technique. In addition, comparative physiological analyses of the static and dynamic myofunctions of the athletes were undertaken to help young athletes perform high-latitude aerial rotation techniques while avoiding damage. The major research hypotheses addressed in this thesis are as follows: First, there would be differences in the 360°, 540°, and 720° rotations of jump inside kicks depending on the injury length of knee joints. Second, there would be differences in the static and dynamic contractile myofunctions depending on the injury length.

2. Materials and Methods

2.1. Participants

The participants of this study were selected from male Wushu national athletes who could perform 360°, 540°, and 720° rotation drills. They were all right-handed and right-footed. Their mean (SD) age was 26.12 (2.84) years old. All of them had suffered knee injuries and were all recovering and returning to training. All players had no cardiovascular problems and were excluded if they had taken any treatment or medication known to affect physical condition or had undergone any major surgery except for a knee joint operation during the one year before the start of this study. The following were also reasons for exclusion: having a history of cerebrovascular disease, impairment of a primary organ system, severe lung disease, cerebral trauma, uncontrolled hypertension, or psychiatric disorder. Twelve players were enrolled in this study. After taking baseline measurements, the group was classified into a group with less than 20 months of injury experience (LESS IG, n = 6) and a group with more than 20 months of injury experience (MORE IG, n = 6). The reason for classifying the group as 20 months was the application of the Cochran–Mantel–Haenszel equation, which expresses the injury risk as the number of knees injured as a percentage of the total number [11]. The subjects of this study were evaluated by a specialist, and athletes with an injury length of 20 months or more were classified as having a high risk of repeated injury, whereas those of less than 20 months were classified as having a moderate injury length. In LESS IG, three patients were diagnosed with right anterior criuciate ligament (ACL) rupture, but recovered after 5, 7, and 8 months of rehabilitation treatment, respectively. On the other hand, one player was diagnosed with a right hamstring sprain and received 5 months of treatment, while the other two players recovered after 18 and 20 months of postoperative rehabilitation treatment for partial rupture of the right knee joint cartilage, respectively. In the MORE IG, two players were diagnosed with a simple right ACL rupture and two players were diagnosed with ACL+PCL (posterior criuciate ligament) complex rupture of the right knee joint, and recovered after 7, 14, 16, and 18 months of rehabilitation treatment, respectively. Meanwhile, one player was diagnosed with a right hamstring rupture and recovered after 12 months of treatment. However, 7 days later, during training, he ruptured again in the same area and had to undergo treatment for another 12 months. The other player was diagnosed with an ACL+PCL complex rupture of the right knee and recovered after 18 months including surgery and rehabilitation. However, this athlete recovered after 18 months of rehabilitation after surgery due to a partial rupture of the right knee joint cartilage during training. Table 1 shows the complete characteristics of the participants.

Table 1. Physical characteristics of the Wushu self-taolu players.

	Groups		Z	ES	p *
	LESS IG	MORE IG			
Age (y)	25.50 ± 2.87	26.75 ± 2.86	−2.683	0.436	0.115
Height (cm)	169.24 ± 3.64	167.80 ± 5.41	1.455	0.312	0.283
Weight (kg)	65.45 ± 6.76	66.64 ± 5.43	−1.327	0.194	0.395
Athletic career (month)	192.00 ± 51.47	200.00 ± 37.69	−0.167	0.177	0.937
Injury period (month)	8.17 ± 4.69	20.00 ± 8.85	−2.330	1.670	0.015

All data represent mean ± standard deviation. * Analyzed using Mann–Whitney U test. LESS IG, a group with less than 20 months of injury experience; MORE IG, a group with more than 20 months of injury experience. ES, effect size; Z, statistical symbol.

2.2. Experimental Design

This study followed the principles of the Declaration of Helsinki and received approval from the institutional ethics committee. Prior to the study, the investigator explained all the procedures to the players who were recruited through advertisements and written informed consent was obtained before enrollment. All players arrived at the research center to sign an informed consent form, complete a self-report questionnaire about their health status included in the physical examination, and to take assessments that consisted of body composition, jump motion, tensiomyography (TMG), and isokinetic moments tests.

2.3. Measurement Methods

2.3.1. Body Composition Measure

An Inbody 230 (Biospace Co., Ltd., Seoul, Korea) analyzer with the bioelectrical impedance analysis method was used for the body composition measurements of all Wushu players. This analyzer is a segmental impedance device that assesses the voltage drop in the upper and lower body. The participants were asked to remove all metal objects and anything else that might interfere with the electric stimuli, including socks, before stepping onto the platform. They were also asked to hold onto the handles and stand still for around 3 min [4,12]. Bioelectrical impedance analysis can track fat mass by conducting high frequency (500~800 KHz) harmless to the human body and using the difference in electrical resistance between adipose tissue and non-adipose tissue. In order to minimize the error, in this study, food intake was abstained 4 h before the test, and alcohol was prohibited 48 h before the test. In addition, exercise was not allowed 12 h before the test. On the other hand, it was necessary to urinate 30 min before the test.

2.3.2. Kinematic Motion Measure

Two-dimensional imaging was performed to analyze the movements of a Wushu jump inside kick at 360°, 540°, and 720°, respectively. Wushu players all took the jump inside kick in a counterclockwise direction. To obtain spatial coordinates for image analysis, a control point frame with three control points was installed with a width of 1 m, a height of 2 m, and a length of 2 m; therefore, all the players' movements were captured, as shown in Figure 1.

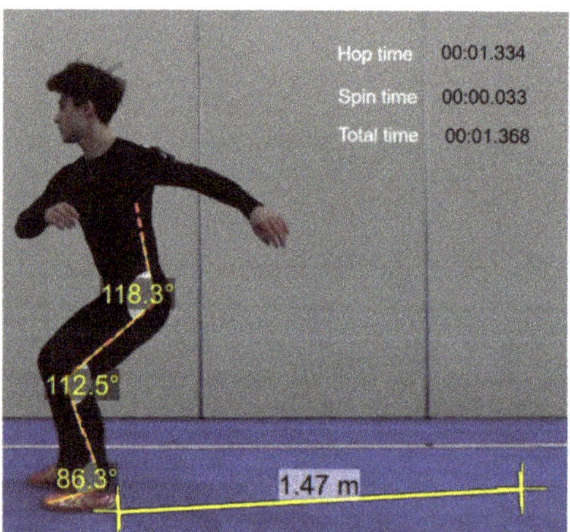

Figure 1. Marker attachment locations. To obtain spatial coordinates for image analysis, a control point frame with three control points was installed. Then, an observer captured images for assessing hop time, spin time, and total time.

The point frame was photographed for about 1 s, after which the control point frame was removed. Image analysis was performed using two high-speed cameras (FDR-AX100, Sony, Japan), and analyzed with Dartfish (Dartfish Live S10 program, Dartfish, Switzerland). A total of two units were installed to cover all the players' performance. For the experiment, each player engaged in warm-up and kicking practice and had three markers attached to each joint area. Then, they performed five drills while being photographed, and the motions that were determined to be the most complete were selected and analyzed.

As shown in Figure 2, five events and four phases were used for the kinematic analysis of three motions of the jump inside kick. Event (E) 1 indicates the preparation stage. E2 indicates the moment of the knee joint of the lower extremity reaching the minimum angle. E3 indicates the moment that the lower leg leaves the ground. E4 indicates the moment the kicking leg reaches the highest point and makes impact with a hand. E5 indicates the moment the foot lands on the ground. Phase (P) 1 indicates the preparation segment from E1 to E2. P2 indicates the take-off segment from E2 to E3. P3 indicates the kicking action from E3 to E4. P4 indicates the turning/spinning and landing segments from E4 to E5.

After recording the entire sequence of events using a camera, the anatomical coordinate points of each frame were digitized.

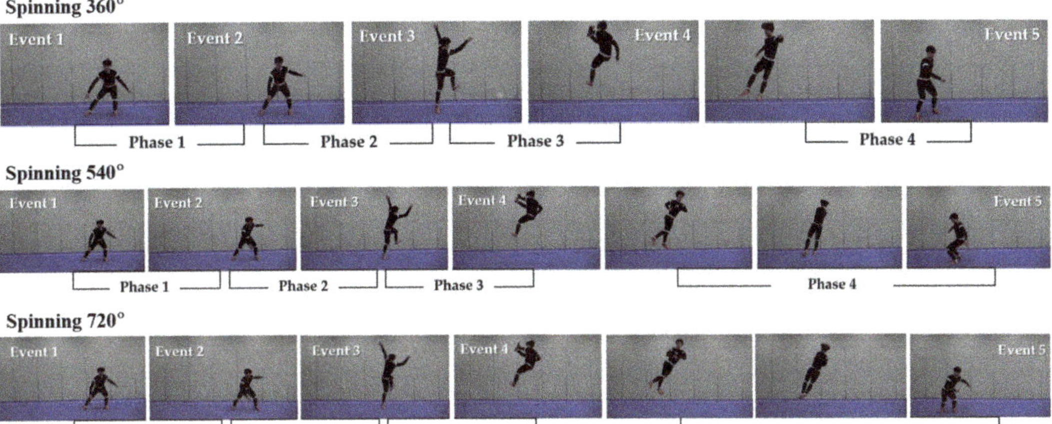

Figure 2. Events and phases in 360°, 540°, and 720° jump inside kicks. Two-dimensional images were performed to analyze the movement of jump inside at 360° (**up**), 540° (**middle**), and 720° (**bottom**). Event (E) 1, preparation stage; E2, moment of knee joint reaching the minimum angle; E3, moment of the lower leg leaving the ground; E4, moment the kicking leg reaching the highest point and making impact with a hand; E5, moment of the foot landing on the ground. Phase (P) 1, preparation segment from E1 to E2; P2, take-off segment from E2 to E3; P3, kicking action from E3 to E4; P4, turning/spinning and landing segments from E4 to E5.

2.3.3. Kinetic Variables Measures

TMG Measure for Static Muscle Contraction

This study employed a TMG device (TMG100, TMG-BMC Ltd., Ljubljana, Slovenia). The extraction of contractile parameters from TMG responses is straightforward and does not require special post-processing of filtering [13,14]. The Wushu players were lying in a prone position on the examination couch. The correct angles in the joints allowing for relaxation of the examined muscles were ensured using guidelines and recommendations of the device manufacturers (GK 40 Panoptik d.o.o., Ljubljana, Slovenia). Two adhesive electrodes stimulating the muscle were placed 2–5 cm apart. The electrodes were placed in a way that did not affect the tendons and allowed the contraction of the particular muscle to be isolated and the simultaneous activation of nearby muscles to be avoided. In the TMG test, the uninjured leg was examined first, and then the leg in the injured area was examined. The placement of the sensor was selected in order to locate the thickest part of the muscle. The sensor was applied to the skin halfway between the electrodes [12]. The electrodes received one 1-millisecond single-phase rectangular pulse from the electrostimulator inducing percutaneous muscle contraction. The pulse power was gradually increased by 10 mA until the maximal contraction reaction was achieved. For minimizing the effects of fatigue, 10 s intervals were taken between the pulses. Typical maximum contraction reactions were recorded between 40 and 80 mA [15]. Displacement–time curve recordings allow muscle contractile properties to be assessed for contraction time (Tc) and maximal displacement (Dm) [13,16,17]. This study measured the Tc and Dm of TMG in the vastus lateralis, vastus medialis, rectus femoris, biceps femoris, gastrocnemius lateralis, gastrocnemius medialis, and tibialis anterior in both legs, as shown in Figure 3.

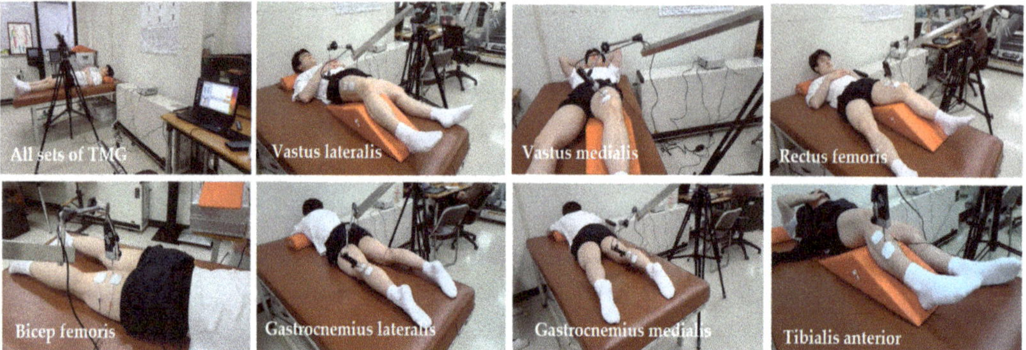

Figure 3. TMG measures' samples. TMG is a technique based on the quantification of radial muscle belly displacement in response to a single electrical stimulus. The measurements are performed in a relaxed position. A digital transducer measured the radial displacement pressed perpendicularly against the skin above the muscle belly in vastus lateralis, vastus medialis, rectus femoris, biceps femoris, gastrocnemius lateralis, gastrocnemius medialis, and tibialis anterior in both legs.

Isokinetic Measure for Dynamic Muscle Contraction

All players were positioned in an isokinetic dynamometer (HUMAC®/NORM™ Testing and Rehabilitation System, CSMi, MA, USA) according to the manufacturer's guidelines, as shown in Figure 4. Testing was performed on the uninjured side first, and then performed on the non-injured side.

Figure 4. Isokinetic moments measures' samples. The knee joint for knee extension and flexion was positioned at 90°. All participants were concentrically tested at 60°/s and 180°/s. The test angles for ankle plantarflexion and dorsiflexion comprised of movement from 50° (plantarflexion) to 20° (dorsiflexion), where all participants were concentrically tested at 30°/s and 120°/s. The test angles for ankle inversion/eversion comprised of movement from 55° (inversion) to 40° (eversion, where all participants were concentrically tested at 60°/s and 90°/s.

All players for the knee extension/flexion test were submitted to a warm-up program before the test. Each participant was placed in the equipment's adjustable seat. The tested limb was placed and fixed with a Velcro strap on a support over the quadriceps, and the knee joint was positioned at 90°. After the participant was positioned and uniformly stabilized, the participant's leg was statically weighed to provide for gravity compensation. Each participant was concentrically tested at 60°/s and at 180°/s. The range of motion (RoM) of extension/flexion was at 0° and 90°. Players then performed 4 maximal warm-up repetitions and 5 maximal test repetitions for evaluating peak torque (Pt) and work per repetition (Wr) at 60°/s, and performed 15 maximal test repetitions for evaluating Pt, fatigue index (Fi), and total work (Tw) at 180°/s. The rest time between the two angular speeds was 60 s.

The players also took the ankle inversion/eversion tests [17]. Each participant was placed in an adjustable seat. The tested limb was placed and fixed with a strap on a support under the gastrocnemius, and the knee was positioned at 30° of flexion. The foot was placed on the inversion/eversion apparatus and fixed with two straps. The axes of the ankle were positioned according to the placement proposed by research [18]. The trunk was stabilized with constraining straps, and an extra strap was used to stabilize the hip at 80° flexion. The arms and the lower limb that was not being tested were placed in a resting position. Two RoM targets consisting of plastic markers were placed at the level of the footplate to facilitate inversion/eversion movements [18]. The players performed 4 submaximal practices at each test speed (60°/s and 90°/s); each test comprised of movement from ankle eversion (40°) to inversion (55°). The measured results of isokinetic moments at 60°/s were analyzed by Pt and Wr. The measures at 90°/s were analyzed by Pt, Fi, and Tw. A rest period of 60 s was given between the tests.

For the ankle plantarflexion/dorsiflexion test, the players were placed in the prone position with the knee placed in full extension and stabilized at the level of the distal thigh with straps. The testing side was placed on the footplate attachment and the lever arm of the dynamometer was aligned with the foot. The motor axis was positioned against the lateral malleolus. The RoM test for both plantar- and dorsiflexions was set during the first test. The players performed 4 submaximal practices at each test speed (30°/s and 120°/s); each cycle comprised of movements from ankle plantarflexion (50°) to dorsiflexion (20°). In the concentric test mode, the players were instructed to push the lever arm along the full angular sector in both directions [19,20]. Both ankles of each participant were assessed and the same rest time was given. The measured isokinetic moments at 30°/s were analyzed by Pt and Wr. The measures at 120°/s were analyzed by Pt, Fi, and Tw.

2.4. Data Process and Statistical Analyses

The sample size was calculated using G Power Software version 3.1.9.7 [21]. The necessary sample size of 12 subjects was calculated from data to achieve a power of 0.40 and an effect size of 0.90, with an α level of 0.05. All data were reported as mean ± standard deviation and carried out using the IBM® SPSS® Statistics software (version 22.0. IBM Corporation; Armonk, NY, USA). The distribution of all data was checked using the Shapiro–Wilk test. Due to non-normally distributed data, a non-parametric Mann–Whitney U test was conducted. Effect sizes were determined by converting partial eta-squared (ES) to Cohen's d [22]. For all analyses, the significance level was set at $p \leq 0.05$.

3. Results

3.1. Analysis of Demographic Variables

As shown in Table 1, although the injury period was a significant difference between the groups, there were no significant differences in the remaining variables of the two groups. The demographic variables of this study indicated the homogeneity of the subjects.

3.2. Kinematic Analysis of Jump Inside Kick

3.2.1. Comparisons of Event Time and Phase Time of Jump Inside Kick

Table 2 shows that the total event time from E1 to E5 at 360° in the LESS IG appeared to be 0.04 s longer than that of the MORE IG. On the other hand, at 540°, the total event time from E1 to E5 in the LESS IG was 0.27 s shorter than that of the MORE IG, indicating that the ability of the LESS IG to rotate in the air was faster than that of the MORE IG. These results were similar at 720°. The results show that the event times of rotation in the LESS IG was significantly faster than those of the MORE IG for 540° and 720° jump inside kicks.

Table 2. Event time and phase time of jump inside kick.

		Groups			
		LESS IG	MORE IG	Z	ES
Event 360°	Event 1 (s)	0.00 ± 0.00	0.00 ± 0.00	-	-
	Event 2 (s)	0.20 ± 0.04	0.20 ± 0.01	-	0.00
	Event 3 (s)	0.42 ± 0.02	0.40 ± 0.10	−1.000	0.277
	Event 4 (s)	0.70 ± 0.00	0.68 ± 0.02	−1.000	0.277
	Event 5 (s)	1.07 ± 0.00	1.07 ± 0.00	-	-
	Total (s)	2.39 ± 0.07	2.35 ± 0.02	−0.775	0.777
Event 540°	Event 1 (s)	0.00 ± 0.00	0.00 ± 0.00	-	-
	Event 2 (s)	0.18 ± 0.03	0.23 ± 0.00	−1.633	0.777
	Event 3 (s)	0.38 ± 0.02	0.45 ± 0.02	−1.549	3.500
	Event 4 (s)	0.62 ± 0.02	0.72 ± 0.02	−2.549 *	5.000
	Event 5 (s)	1.05 ± 0.03	1.10 ± 0.00	−1.633	5.000
	Total (s)	2.23 ± 0.01	2.50 ± 0.05	−2.562 *	7.488
Event 720°	Event 1 (s)	0.00 ± 0.00	0.00 ± 0.00	-	-
	Event 2 (s)	0.17 ± 0.05	0.17 ± 0.00	-	7.488
	Event 3 (s)	0.36 ± 0.00	0.40 ± 0.00	−1.633	7.488
	Event 4 (s)	0.62 ± 0.02	0.63 ± 0.00	-	7.488
	Event 5 (s)	1.07 ± 0.05	1.05 ± 0.02	0.001	0.525
	Total (s)	2.11 ± 0.11	2.75 ± 0.02	−2.532 *	8.095
Phase 360°	Phase 1 (s)	0.12 ± 0.16	0.20 ± 0.00	0.001	8.095
	Phase 2 (s)	0.19 ± 0.02	0.20 ± 0.00	-	8.095
	Phase 3 (s)	0.34 ± 0.09	0.32 ± 0.02	0.001	0.306
	Phase 4 (s)	0.54 ± 0.23	0.39 ± 0.02	−0.408	0.918
	Total (s)	1.17 ± 0.14	1.10 ± 0.04	−1.608	0.679
Phase 540°	Phase 1 (s)	0.10 ± 0.14	0.23 ± 0.00	−2.633 *	0.679
	Phase 2 (s)	0.20 ± 0.05	0.22 ± 0.03	−0.775	0.485
	Phase 3 (s)	0.32 ± 0.12	0.27 ± 0.01	0.001	0.587
	Phase 4 (s)	0.55 ± 0.11	0.39 ± 0.02	−2.549 *	2.023
	Total (s)	1.16 ± 0.04	1.10 ± 0.00	−1.633	2.023
Phase 720°	Phase 1 (s)	0.10 ± 0.14	0.17 ± 0.00	0.001	27.457
	Phase 2 (s)	0.15 ± 0.02	0.23 ± 0.00	−2.633 *	27.457
	Phase 3 (s)	0.32 ± 0.06	0.24 ± 0.00	−2.651	27.457
	Phase 4 (s)	0.14 ± 0.09	0.37 ± 0.05	−2.549 *	3.159
	Total (s)	1.10 ± 0.01	2.09 ± 0.05	−2.862 *	27.457

All data represent mean ± standard deviation. LESS IG, a group with less than 20 months of injury experience; MORE IG, a group with more than 20 months of injury experience. ES, effect size; *, $p < 0.05$.

Table 2 also shows that the total phase times at 360° and at 540° did not show a significant difference between the two groups, but at 720° rotation, it was confirmed that the LESS IG was significantly faster than the MORE IG.

3.2.2. Comparisons of Distance and Time of Jump Inside Kick

Table 3 represents the distance and time from take-off to landing between the LESS IG and the MORE IG. For the total distance covered in the 360° jump inside kick, the LESS IG was shorter than the MORE IG. This trend was similar in 540° and 720° rotation, and it can be observed that there is a statistically significant difference between the groups.

Table 3. Distance and time required by events of jump inside kick.

			Groups			
			LESS IG	MORE IG	Z	ES
360°	DPB	d (m)	2.19 ± 0.99	2.94 ± 0.42	−0.775	0.986
		t (s)	1.72 ± 0.55	2.59 ± 0.73	−0.786	1.346
	DSL	d (m)	0.95 ± 0.74	1.66 ± 0.68	−0.775	0.999
		t (s)	0.98 ± 0.21	0.77 ± 0.14	−0.775	1.176
	TDPL	d (m)	3.14 ± 1.73	4.60 ± 1.10	−2.973 *	1.007
		t (s)	2.70 ± 0.33	3.35 ± 0.59	−2.775 *	1.359
540°	DPB	d (m)	2.51 ± 1.00	2.89 ± 0.11	0.001	0.534
		t (s)	1.49 ± 0.45	1.82 ± 0.12	−0.775	1.002
	DSL	d (m)	1.14 ± 0.70	1.34 ± 0.01	0.001	0.404
		t (s)	0.98 ± 0.16	0.75 ± 0.17	−1.225	1.393
	TDPL	d (m)	3.04 ± 1.70	4.23 ± 0.11	2.589 *	0.987
		t (s)	2.47 ± 0.28	3.57 ± 0.28	−2.775 *	3.928
720°	DPB	d (m)	2.87 ± 0.78	3.36 ± 0.44	−2.917 *	0.773
		t (s)	1.95 ± 0.21	2.27 ± 0.42	−2.768 *	0.963
	DSL	d (m)	1.28 ± 0.96	1.44 ± 0.41	−1.091	0.216
		t (s)	1.10 ± 0.00	0.90 ± 0.00	−1.633	0.216
	TDPL	d (m)	3.15 ± 1.75	4.80 ± 0.85	−2.675 *	1.199
		t (s)	3.05 ± 0.21	4.17 ± 0.42	−2.901 *	3.373

All data represent mean ± standard deviation (SD). LESS IG, a group with less than 20 months of injury experience; MORE IG, a group with more than 20 months of injury experience; ES, effect size; d, distance; t, time. DPB, Distance from prior jump to before jump; DSL, Distance from spinning to landing; TDPL, Total distance from take-off to landing. * $p < 0.05$.

3.3. Kinetic Analysis of Jump Inside Kick

3.3.1. Analysis of Static Muscle Contractions in Lower Legs

Table 4 shows that there was a significant difference in Tc between the LESS IG and the MORE IG. The Tcs of the right gastrocnemius lateralis, right gastrocnemius medialis, and right tibialis anterior of the LESS IG were significantly higher than those of the MORE IG. Meanwhile, the Tcs of left rectus femoris, left tibialis anterior and left vastus medialis of the LESS IG were significantly higher than those of the MORE IG. Dm also showed a similar tendency to Tc. This tendency can be seen as a phenomenon that occurs because the Wushu athletes must rotate to the left with the center point on their left leg in order to perform a jump inside kick.

3.3.2. Analysis of Dynamic Muscle Contractions in Isokinetic Low Angular Speed

Table 5 shows that there were significant differences in the isokinetic knee extensor and flexor at 60°/s between the LESS IG and the MORE IG. Specifically, Pt and Wr in the left extensor and flexor of the LESS IG were significantly higher than those of the MORE IG. A similar tendency was observed in the ankle invertor/evertor, although there were no significant differences between both groups. In the ankle plantarflexion/dorsiflexion test at 30°/s, the Pt and Wr in the right plantarflexor of the LESS IG were significantly higher than those of the MORE IG.

Table 4. Comparison of contraction time from TMG of lower legs.

Measured Muscles	TMG Variables	Groups		Z	ES
		LESS IG	MORE IG		
Biceps Femoris	Tc (ms) of Lt	21.42 ± 0.13	23.63 ± 4.57	−0.684	0.683
	Dm (mm) of Lt	3.27 ± 2.69	2.31 ± 0.67	0.493	0.489
	Tc (ms) of Rt	22.68 ± 3.90	19.61 ± 6.99	0.543	0.542
	Dm (mm) of Rt	1.09 ± 0.42	4.85 ± 0.40	−5.425 **	9.168
Gastrocnemius Lateralis	Tc (ms) of Lt	26.82 ± 11.19	25.80 ± 3.06	0.125	0.124
	Dm (mm) of Lt	2.11 ± 0.74	4.25 ± 0.62	−4.504 **	3.134
	Tc (ms) of Rt	32.69 ± 3.76	16.23 ± 24.13	2.957 *	0.953
	Dm (mm) of Rt	1.87 ± 1.73	5.79 ± 1.56	−2.784 *	2.379
Gastrocnemius Medialis	Tc (ms) of Lt	24.04 ± 11.82	20.55 ± 0.03	0.417	0.417
	Dm (mm) of Lt	1.65 ± 0.62	1.92 ± 0.08	−0.608	0.610
	Tc (ms) of Rt	37.17 ± 26.65	15.82 ± 2.61	3.428 *	1.127
	Dm (mm) of Rt	0.72 ± 0.61	0.96 ± 0.76	−1.115	0.348
Rectus Femoris	Tc (ms) of Lt	21.99 ± 0.27	13.74 ± 0.25	4.418 **	31.707
	Dm (mm) of Lt	4.37 ± 1.15	8.80 ± 1.78	−2.376 *	2.956
	Tc (ms) of Rt	18.33 ± 5.43	18.75 ± 2.01	−0.103	0.102
	Dm (mm) of Rt	4.34 ± 1.48	4.74 ± 0.06	−0.376	0.381
Tibialis Anterior	Tc (ms) of Lt	29.01 ± 2.84	17.28 ± 6.06	2.746 *	2.478
	Dm (mm) of Lt	0.83 ± 0.42	1.03 ± 0.37	−0.504	0.505
	Tc (ms) of Rt	49.17 ± 0.49	22.86 ± 42.41	2.990 *	0.877
	Dm (mm) of Rt	2.07 ± 0.76	2.11 ± 1.73	−0.030	0.029
Vastus Lateralis	Tc (ms) of Lt	22.73 ± 5.56	20.94 ± 1.07	0.301	0.447
	Dm (mm) of Lt	3.20 ± 2.88	4.92 ± 1.66	−0.732	0.731
	Tc (ms) of Rt	19.00 ± 10.92	18.78 ± 2.84	0.348	0.027
	Dm (mm) of Rt	3.43 ± 4.54	3.41 ± 0.42	0.006	0.006
Vastus Medialis	Tc (ms) of Lt	29.94 ± 1.75	17.56 ± 3.67	2.916 *	4.306
	Dm (mm) of Lt	2.26 ± 2.75	5.56 ± 0.25	−1.383	1.690
	Tc (ms) of Rt	22.74 ± 6.81	25.14 ± 3.85	−0.434	0.433
	Dm (mm) of Rt	4.46 ± 0.56	4.70 ± 0.86	−0.332	0.330

All data represent mean ± standard deviation (SD). LESS IG, a group with less than 20 months of injury experience; MORE IG, a group with more than 20 months of injury experience; ES, effect size; TMG, tensiomyography; Tc, contraction time; Dm, maximal displacement. *, $p < 0.05$; **, $p < 0.01$.

Table 5. Comparison results of isokinetic knee extension/flexion at 60°/s, ankle inversion/eversion at 60°/s, and ankle plantarflexion/dorsiflexion at 30°/s.

Measured Muscles	Pt/Wr of Sides	Groups		Z	ES
		LESS IG	MORE IG		
Knee Extensor	Pt (Nm) of Rt	194.00 ± 69.30	145.50 ± 13.44	0.972	0.971
	Pt (Nm) of Lt	220.00 ± 29.70	156.50 ± 21.92	2.433 *	2.432
	Wr (Nm) of Rt	174.50 ± 40.31	145.50 ± 10.61	0.984	0.983
	Wr (Nm) of Lt	208.00 ± 14.14	151.00 ± 29.70	2.451 *	2.450
Knee Flexor	Pt (Nm) of Rt	152.00 ± 48.08	105.50 ± 7.78	1.350	1.350
	Pt (Nm) of Lt	156.50 ± 4.95	96.00 ± 24.04	3.486 **	3.485
	Wr (Nm) of Rt	158.00 ± 8.49	125.00 ± 24.04	1.831	1.830
	Wr (Nm) of Lt	167.00 ± 22.63	102.00 ± 29.70	2.462 *	2.461
Ankle Invertor	Pt (Nm) of Rt	18.00 ± 0.01	18.50 ± 0.71	−1.000	0.995
	Pt (Nm) of Lt	23.00 ± 5.66	16.00 ± 2.83	1.565	1.564
	Wr (Nm) of Rt	10.00 ± 1.41	11.00 ± 0.01	−1.000	1.002
	Wr (Nm) of Lt	15.50 ± 6.36	11.00 ± 0.01	1.000	1.000

Table 5. *Cont.*

Measured Muscles	Pt/Wr of Sides	Groups		Z	ES
		LESS IG	MORE IG		
Ankle Evertor	Pt (Nm) of Rt	17.00 ± 2.83	16.50 ± 3.54	0.156	0.156
	Pt (Nm) of Lt	19.00 ± 0.01	17.50 ± 2.12	1.000	1.000
	Wr (Nm) of Rt	11.00 ± 0.01	10.50 ± 2.12	0.333	0.333
	Wr (Nm) of Lt	14.50 ± 4.95	13.00 ± 1.41	0.412	0.412
Ankle Plantarflexor	Pt (Nm) of Rt	123.00 ± 14.14	104.00 ± 8.49	2.629 *	1.629
	Pt (Nm) of Lt	101.50 ± 7.78	105.00 ± 8.49	−0.430	0.429
	Wr (Nm) of Rt	71.00 ± 2.83	43.50 ± 13.44	2.833 *	2.831
	Wr (Nm) of Lt	50.50 ± 4.95	47.50 ± 2.12	0.788	0.787
Ankle Dorsiflexor	Pt (Nm) of Rt	37.00 ± 5.66	39.00 ± 5.66	−0.354	0.353
	Pt (Nm) of Lt	39.50 ± 3.54	33.50 ± 4.95	1.395	1.394
	Wr (Nm) of Rt	24.50 ± 7.78	17.00 ± 1.41	1.342	1.341
	Wr (Nm) of Lt	24.00 ± 5.66	15.00 ± 4.24	1.800	1.799

All data represent mean ± standard deviation. LESS IG, a group with less than 20 months of injury experience; MORE IG, a group with more than 20 months of injury experience; ES, effect size; Pt, peak torque; Wr, work per repetition; Rt, right; Lt, left. *, $p < 0.05$; **, $p < 0.01$.

3.3.3. Analysis of Dynamic Muscle Contractions in Isokinetic Moderate to High Angular Speed

Table 6 shows that there were significant differences in isokinetic knee extensor and flexor at 180°/s between the LESS IG and the MORE IG. The knee extensor that serves as a crutch when rotating to the left was significantly higher in the left leg of the LESS IG, whereas the knee flexor that acts as a crutch when performing a jump was significantly higher in the right leg. A similar tendency was also observed in the isokinetic ankle invertor/evertor and ankle plantarflexor/dorsiflexor.

Table 6. Comparison results of isokinetic knee extension/flexion at 180°/s, ankle inversion/eversion at 90°/s, and ankle plantarflexion/dorsiflexion at 120°/s.

Measured Muscles	Pt/Fi/Tw of Sides	Groups		Z	ES
		LESS IG	MORE IG		
Knee Extensor	Pt (Nm) of Rt	131.00 ± 35.36	110.50 ± 6.36	0.807	0.806
	Pt (Nm) of Lt	147.50 ± 16.26	95.50 ± 37.48	2.800 *	1.799
	Fi of Rt	10.00 ± 9.90	8.50 ± 0.71	0.214	0.213
	Fi of Lt	20.50 ± 7.78	−2.50 ± 20.51	3.483 **	1.482
	Tw (Nm) of Rt	1661.00 ± 65.05	1652.50 ± 133.64	0.081	0.080
	Tw (Nm) of Lt	1824.50 ± 65.76	1350.50 ± 525.38	2.566 *	1.266
Knee Flexor	Pt (Nm) of Rt	110.50 ± 4.95	83.50 ± 6.36	2.736 *	4.737
	Pt (Nm) of Lt	122.50 ± 10.61	68.50 ± 27.58	3.585 **	2.584
	Fi of Rt	9.50 ± 2.12	9.00 ± 19.80	0.036	0.035
	Fi of Lt	14.00 ± 1.41	−8.00 ± 46.67	1.666	0.666
	Tw (Nm) of Rt	1504.00 ± 333.75	1266.50 ± 125.16	2.942 *	0.942
	Tw (Nm) of Lt	1614.50 ± 365.57	1028.00 ± 258.80	2.852 *	1.851
Ankle Invertor	Pt (Nm) of Rt	16.00 ± 0.01	14.50 ± 0.71	1.256	2.987
	Pt (Nm) of Lt	16.00 ± 2.83	15.00 ± 4.24	0.277	0.274
	Fi of Rt	21.00 ± 18.38	4.50 ± 13.44	3.025 **	1.024
	Fi of Lt	5.00 ± 1.41	21.50 ± 23.33	−3.998 **	0.998
	Tw (Nm) of Rt	124.00 ± 19.80	135.00 ± 19.80	−0.556	0.555
	Tw (Nm) of Lt	168.50 ± 75.66	125.50 ± 6.36	1.801	0.800

Table 6. *Cont.*

Measured Muscles	Pt/Fi/Tw of Sides	Groups		Z	ES
		LESS IG	MORE IG		
Ankle Evertor	Pt (Nm) of Rt	16.00 ± 2.83	13.00 ± 2.83	1.061	1.060
	Pt (Nm) of Lt	17.00 ± 1.41	16.50 ± 3.54	0.186	0.185
	Fi of Rt	28.00 ± 4.24	19.50 ± 4.95	1.844	1.844
	Fi of Lt	24.00 ± 8.49	38.00 ± 2.83	−1.972	2.212
	Tw (Nm) of Rt	122.00 ± 19.80	106.50 ± 12.02	0.946	0.944
	Tw (Nm) of Lt	169.50 ± 61.52	126.00 ± 1.41	2.030	0.999
Ankle Plantarflexor	Pt (Nm) of Rt	72.00 ± 11.31	54.50 ± 7.78	1.803	0.021
	Pt (Nm) of Lt	62.00 ± 8.49	52.50 ± 4.95	1.368	1.367
	Fi of Rt	17.00 ± 18.38	21.00 ± 1.41	−0.307	0.306
	Fi of Lt	9.50 ± 20.51	21.50 ± 7.78	−0.774	0.773
	Tw (Nm) of Rt	594.00 ± 48.08	322.50 ± 137.89	4.629 **	2.629
	Tw (Nm) of Lt	520.50 ± 159.10	372.50 ± 126.57	2.730 *	1.029
Ankle Dorsiflexor	Pt (Nm) of Rt	22.50 ± 0.71	25.00 ± 4.24	−0.822	1.151
	Pt (Nm) of Lt	26.00 ± 0.01	23.50 ± 0.71	5.000	2.987
	Fi of Rt	18.00 ± 5.66	22.00 ± 29.70	−0.187	0.233
	Fi of Lt	14.00 ± 11.31	20.00 ± 7.07	−0.636	0.742
	Tw (Nm) of Rt	186.50 ± 17.68	153.50 ± 36.06	1.866	1.174
	Tw (Nm) of Lt	198.50 ± 58.69	152.50 ± 33.23	1.965	0.964

All data represent mean ± standard deviation. LESS IG, a group with less than 20 months of injury experience; MORE IG, a group with more than 20 months of injury experience; ES, effect size; Pt, peak torque; Fi, fatigue index; Tw, total work; Rt, right; Lt, left. *, $p < 0.05$; **, $p < 0.01$.

4. Discussion

The findings of this study revealed that when comparing the Wushu self-taolu athletes who have had different injury periods, there was no significant difference in body composition and the 360° jump inside kick. However, there were significant differences between the groups for the 540° and 720° jump inside kicks. In addition, in the event and phase time related to rotational time, there were no differences between the groups for the 360° jump inside kick, but there were significant differences between the groups at the higher difficulty levels. In other words, the LESS IG had a shorter jump distance, but faster rotational time, although there was no significant difference in body composition compared to the MORE IG. Specifically, this study found that there were significant differences between the groups in the static and dynamic myofunctions between the LESS IG and the MORE IG.

When the jump inside kick was performed, the time in mid-air for the LESS IG was longer than that of the MORE IG prior to completing the inside kick. In this study, it is thought that the LESS IG efficiently performed the jump with sufficient space and elapsed time during the inside kick motions at 360°, 540°, and 720°. The distance and time from take-off to landing between the LESS IG and the MORE IG also showed that the LESS IG had a shorter jump distance, but a faster rotational time; such a fast rotation time must be supported by the myofunction capable of exerting force. The jump inside kick is a technique that starts with running from a distance, jumping counterclockwise while jumping (Phases two to three), and kicking with the inside surface of the right foot while jumping (Event four). In other words, touching the sole of the right foot during the three types of jump inside kicks occurred close to the end of Event four. For a better jump front kick, an athlete needs to be able to jump vertically. For this purpose, the forward speed just before hopping should be fast; during the hopping period, it is important to shift this forward speed vertically [6]. The better the athlete is, the shorter the hop time in the run before and after the jump is; the shorter the intersection between the vertical and horizontal speeds in the section immediately before and after the jump is, the more efficient the performance is [23].

At international competitions, there are many foreign Wushu athletes who perform at C level, and the results are excellent. In Korea, there are only few athletes who can perform at C level. Self-taolu athletes must perform advanced skills in order to excel in the Asian Games and World Championships. Therefore, this study intended to present the analysis and application method for performing high-difficulty techniques. Recently, many quantified data and instruction models have been proposed through these analyses. Most studies have only analyzed difficulty levels for A and B, such as the 360° and 540° jump inside kicks, while research on C-level techniques, such as the 720° jump inside kick, is rare. When performing the three types of jump inside kicks, the LESS IG showed superior time and distance compared to the MORE IG. In this regard, Leporace et al. reported that it is desirable to increase the vertical velocity in the jump motion, and to perform the motion using the rotation of the lower joint on the long axis [6]. This study found that although there were no differences at 360° between both groups, there were significant differences in the technical skills in the 540° and 720° jump inside kicks between the groups. That is, for the higher-ranking levels of Wushu athletes in Korea, more technical training is needed at over 360° spinning skill such as a shorter spinning time [1,4,24].

In order to perform the higher levels of difficulty for the jump inside kick, the muscle contractility of the lower extremities is very important [6,25]. In this regard, this study investigated the static muscle contractility of the lower extremity in the Wushu self-taolu players. In the results of this study, Dm in the right biceps femoris of the LESS IG was significantly lower than that of the MORE IG. This indicates a high muscle tone expressed in the hamstrings of the LESS IG may be a concern in the future due to excessive stiffness. It is also evidence that the training related to jumping is repeated periodically. In addition, the high Tc and low Dm of the right gastrocnemius lateralis of the LESS IG showed a significant difference with the values of the MORE IG, and gastrocnemius medialis of the LESS IG showed similar results. On the other hand, compared to the MORE IG, the low Dm of the left gastrocnemius lateralis of the LESS IG, the high Tc and low Dm of the left rectus femoris, and the high Tc of the left tibialis anterior and vastus medialis can be interpreted to mean that the LESS IG is better trained or rehabilitated for jumping and rotating compared to the MORE IG. In this regard, García-Manso et al. reported that the post-activation potentiation after strength exercises was responsible for the observed changes in muscle response at the end of the first set [26]. This mechanism becomes significant with changes in increased Tc and decreased Dm. Several researchers also suggested that a low Dm indicates a high muscle tone and excessive stiffness in the muscle structures, whereas an elevated Dm indicates a lack of muscle tone or the appearance of muscle fatigue [12,13,26]. Considering that Dm varies depending on the load of strength training, the recovery time between repetitions, and the type of muscle contraction, it can be estimated that the LESS IG analyzed in this study had a higher training volume than the MORE IG. The normal curve from TMG has a steep shape, and Tc appears at short intervals [27]. Characteristically, the shape of the overall curve appears to collapse after the injury, which is due to the fact that muscle contraction does not proceed normally and rapidly [26,27]. The two parameters enable the muscular composition of the studied muscular groups to be increased [28], correlated with the increase in the contraction time, and the decrease in the muscular displacement amplitude [12,29]. These parameters have normal average values for Tc, namely, 32.83 ms, and for all muscular groups, the average Dm value is 8.17 mm [29,30]. In the static muscle contractions in this study, there was no significant difference in Tc between the LESS IG and the MORE IG, but the Tc of the left rectus femoris, the Dm of the right biceps femoris, and the Dm of the gastrocnemius lateralis in the LESS IG were significantly higher than those in the MORE IG. The Tc of the right biceps femoris in the LESS IG was also significantly higher than the MORE IG. In other words, the left rectus femoris, right biceps femoris, and right gastrocnemius lateralis are all muscle groups that are required for leaping and kicking, which are significantly higher in the LESS IG than the MORE IG. In particular, the Tc of right biceps femoris was very high in the LESS IG, and it was found to satisfy the hypothesis of this study. It was also found that Dm was most effective in detecting muscle hypertrophy, and that muscle stiffness secondary to muscle hypertrophy could induce a decrease in the Dm of Wushu self-taolu athletes. These results show the fact that when the injury period is short when landing after an inside jump, it is possible to show a little better performance. In other words, as a Wushu athlete, in order to become a superior athlete, it is necessary to have some level of static muscle function in the lower extremities, and this result is considered to be the only way to prevent injuries [31].

Some of the contractile parameters measured by TMG have been found to correlate with muscle peak torque [32], and with the spatial distribution of fiber types in human muscles [14,16,28]. In addition to this relation, the dynamic phenomenon in sports shows that the role of dynamic myofunction as well as static myofunction is very important [33]. From this point of view, this study also investigated the static as well as the dynamic myofunctions, and the following results were observed. The knee extensor muscles that include the vastus medialis, vastus lateralis, and rectus femoris, fixed the knee joint upon landing after the jump. These muscle groups did not differ significantly between the groups in this study. However, the biceps femoris, an important knee flexor muscle at the time of the jump, appeared very predominantly in the LESS IG. In examining the dynamic muscle contractility, the moments of isokinetic low angular speeds in the LESS IG were higher than those of the MORE IG. Specifically, the Pt and Wr of the left knee extensor and flexor of the LESS IG were significantly higher than those of the MORE IG. Through this myofunction of the thigh, explosive jumps are made, and it helps to maintain balance even when landing after a mid-air rotation. This phenomenon is thought to be the development of extensor and flexor muscles in the center of the knee joint, which must withstand the force applied to the left lower extremity just before the jump when athletes need to jump in the left direction. Moreover, the right ankle plantarflexor, which triggers the jumping force to the left, was significantly higher in the LESS IG than in the MORE IG, which is interpreted to suggest the importance of the right plantarflexor that triggers the jumping time to the left. These results seem to support the results shown in the TMG test presented in Table 4. That is, the Tc of the right gastrocnemius lateralis was 32.69 ± 3.76 ms in the LESS IG, while it was 16.23 ± 24.13 ms in the MORE IG, which is consistent with the result of the dynamic contractility of the right ankle plantarflexor.

As shown in Table 6, the moments of isokinetic moderate to high angular speeds are shown similarly to the moments of low angular speed, although it shows specific results. That is, the Pt of the left knee extensor of the LESS IG was significantly higher than that of the MORE IG, and the Tw also showed a similar result, suggesting that the role of the left lower extremity is important before jumping and during landing [4,6,33]. The Fi in the LESS IG is significantly higher than in the MORE IG, which means that muscle fatigue in the LESS IG is high in the left thigh, confirming the meaning of the results once again [31]. On the other hand, unlike the extensor of the knee, Pt and Tw of the knee flexor of the LESS IG were significantly higher than those of the MORE IG on both the left and right sides, whereas there was no difference in Fi. This result shows that the knee flexor should be excellent in bending the knee joint before jumping and absorbing the shock during landing, while the muscle fatigue should be low. Meanwhile, the Tw of the right ankle plantarflexor in the LESS IG was 594.00 ± 48.08 Nm, whereas that in the MORE IG was 322.50 ± 137.89 Nm, which shows a significant difference between both groups. The Tw of the left ankle plantarflexor in the LESS IG was 520.50 ± 159.10 Nm, whereas that of the MORE IG was 372.50 ± 126.57 Nm, which shows a significant difference between both groups. This result shows that when performing the Wushu jump inside kick, the ankle plantarflexor for extending the ankle should be able to perform myofunction without fatigue for a long time. This study confirmed that the excellent rotational skills of jump inside kicks in the Wushu self-taolu athletes come from the well-equipped static and dynamic muscle contractile properties while minimizing the injury frequency or length. Ultimately, this study found that when Wushu players rotate 360, 540, and 720 degrees in the air, the rotation speed varies according to the three target rotation angles, and there is a significant difference according to the length of the injury. However, when the length of the injuries of the athletes is long, the muscle function of the lower extremities to cause rotation in the air is low, and moreover, when the number of rotations is high, it is more evident. These results suggest that even if an athlete is injured, treatment should be carried out quickly so that the frequency or length of the injury does not occur as often.

5. Conclusions

First, the event time, phase time, and distance/time for 540° and 720° jump inside kicks of the LESS IG were higher than those of the MORE IG. Second, there were significant differences in the Tc and Dm between the LESS IG and the MORE IG. The left lower extremity of the LESS IG was well developed. The right biceps femoris was significantly higher in the LESS IG than in the MORE IG. This tendency was similarly observed at isokinetic low to high angular speeds. It can be interpreted that the static and dynamic muscle contractile properties required for the jump inside kicks were significantly higher in the less experienced injury of lower legs in the Wushu athletes. This study confirmed that in order to be a non-injured self-taolu athlete, it is necessary to develop myofunctions of the knee extensor, knee flexor, and ankle plantarflexor for jump inside kicks. In other words, in preparation for injury to the lower extremities that may occur during jumping and landing, muscle function training should be continued. Additionally, it can be concluded that it is desirable to train and to rehabilitate accordingly because there is less torsion of the lower extremities upon landing after a high air jump. However, this study had a small sample size due to the limited number of Wushu players in Korea. In addition, healthy Wushu players were not included in the study. Therefore, there are limitations in being able to generalize these results to other populations and types of competitions.

Author Contributions: J.-Y.C. and Y.-S.J. conceived the idea. S.P. and H.-S.L. developed the background and performed the calibration of different devices used in the tests. J.-Y.C. and H.-S.L. verified the methods section. All authors discussed the results and contributed to the final manuscript. H.-S.L. and J.-Y.C. performed the tests. Y.-S.J. wrote the manuscript with support from S.P. All authors contributed to the final version of the manuscript. S.P. and Y.-S.J. contributed to the interpretation of the results and data analysis, and they drafted the manuscript and designed the figures and tables. All authors provided critical feedback and helped shape the research, analysis, and manuscript.

Funding: This research was not supported by any grant.

Institutional Review Board Statement: The study was conducted according to the guidelines of the Declaration of Helsinki and approved by the Institutional Review Board of Sahmyook University (SYUIRB2015-008).

Informed Consent Statement: Informed consent was obtained from all subjects involved in the study.

Acknowledgments: The authors wish to thank the participants of this study.

Conflicts of Interest: The authors declare no competing interests.

References

1. Lee, S.B.; Hong, J.H.; Lee, T.S. Analysis of physical activities in Wu-Shu training. In Proceedings of the 29th Annual International Conference of the IEEE Engineering in Medicine and Biology Society, Lyon, France, 23–26 August 2007; pp. 632–635. [CrossRef]
2. McAnulty, S.; McAnulty, L.; Collier, S.; Souza-Junior, T.P.; McBride, J. Tai Chi and Kung-Fu practice maintains physical performance but not vascular health in young versus old participants. *Phys. Sportsmed.* **2016**, *44*, 184–189. [CrossRef] [PubMed]
3. Artioli, G.G.; Gualano, B.; Franchini, E.; Batista, R.N.; Polacow, V.O.; Lancha, A.H., Jr. Physiological, performance, and nutritional profile of the Brazilian Olympic Wushu (kung-fu) team. *J. Strength Cond. Res.* **2009**, *23*, 20–25. [CrossRef] [PubMed]
4. Cha, J.Y.; Jee, Y.S. Wushu Nanquan training is effective in preventing obesity and improving heart function in youth. *J. Exerc. Rehabil.* **2018**, *14*, 466–472. [CrossRef] [PubMed]
5. Tsang, T.W.; Kohn, M.R.; Chow, C.M.; Singh, M.A.F. Kung fu training improves physical fitness measures in overweight/obese adolescents: The "martial fitness" study. *J. Obes.* **2010**, *2010*, 672751. [CrossRef] [PubMed]
6. Leporace, G.; Praxedes, J.; Pereira, G.R.; Chagas, D.; Pinto, S.; Batista, L.A. Activation of hip and knee muscles during two landing tasks performed by male volleyball athletes. *Rev. Bras. Med. Esporte.* **2011**, *17*, 324–328. [CrossRef]
7. Kawamori, N.; Nosaka, K.; Newton, R.U. Relationships Between Ground Reaction Impulse and Sprint Acceleration Performance in Team Sport Athletes. *J. Strength Cond. Res.* **2013**, *27*, 568–573. [CrossRef] [PubMed]

8. Piazza, M.; Battaglia, C.; Fiorilli, G.; Innocenti, G.; Iuliano, E.; Aquino, G.; Calcagno, G.; Giombini, A.; Cagno, A.D. Effects of resistance training on jumping performance in pre-adolescent rhythmic gymnasts: A randomized controlled study. *Ital. J. Anat. Embryol.* **2014**, *119*, 10–19. [CrossRef]
9. Hooren, B.V.; Teratsias, P.; Hodson-Tole, E.F. Ultrasound imaging to assess skeletal muscle architecture during movements: A systematic review of methods, reliability, and challenges. *J. Appl. Physiol.* **2020**, *128*, 978–999. [CrossRef]
10. Kuan, G.; Roy, J. Goal profiles, mental toughness and its influence on performance outcomes among Wushu athletes. *J. Sports Sci. Med.* **2007**, *6*, 28–33.
11. Mantel, N.; Haenszel, W. Statistical aspects of the analysis of data from retrospective studies of disease. *J. Natl. Cancer Inst.* **1959**, *22*, 719–748.
12. Park, S.; Park, S.; Yoo, J.; Jee, Y.S. Effects of equine riding on static and dynamic mechanical contraction of the thighs and trunk muscles in inactive women. *J. Back Musculoskelet. Rehabil.* **2021**, *34*, 521–535. [CrossRef] [PubMed]
13. Križaj, D.; Šimunič, B.; Žagar, T. Short-term repeatability of parameters extracted from radial displacement of muscle belly. *J. Electromyogr. Kinesiol.* **2008**, *18*, 645–651. [CrossRef]
14. Dahmane, R.; Djorddievic, S.; Šimunic, B.; Valencic, V. Spatial fiber type distribution in normal human muscle histochemical and tensiomyographical evaluation. *J. Biomech.* **2005**, *38*, 2451–2459. [CrossRef] [PubMed]
15. Domaszewski, P.; Pakosz, P.; Konieczny, M.; Bączkowicz, D.; Sadowska-Krępa, E. Caffeine-induced effects on human skeletal muscle contraction time and maximal displacement measured by tensiomyography. *Nutrients* **2021**, *13*, 815. [CrossRef]
16. Sánchez-Sánchez, J.; Bishop, D.; García-Unanue, J.; Ubago-Guisado, E.; Hernando, E.; López-Fernández, J.; Colino, E.; Gallardo, L. Effect of a repeated sprint ability test on the muscle contractile properties in elite futsal players. *Sci. Rep.* **2018**, *8*, 17284. [CrossRef]
17. Dahmane, R.; Valenčič, V.; Knez, N.; Eržen, I. Evaluation of the ability to make non-invasive estimation of muscle contractile properties on the basis of the muscle belly response. *Med. Biol. Eng. Comput.* **2001**, *39*, 51–55. [CrossRef]
18. Kim, J.D.; Oh, H.W.; Lee, J.H.; Cha, J.Y.; Ko, I.G.; Jee, Y.S. The effect of inversion traction on pain sensation, lumbar flexibility and trunk muscles strength in patients with chronic low back pain. *Isokin. Exerc. Sci.* **2013**, *21*, 237–246. [CrossRef]
19. Leslie, M.; Zachaewski, J.; Browne, P. Reliability of torque values for ankle invertors and evertors. *J. Orthop. Sports Phys. Ther.* **1990**, *11*, 612–616. [CrossRef] [PubMed]
20. Olmo, J.; Jato, S.; Benito, J.; Martín, I.; Dvir, Z. Identification of feigned ankle plantar and dorsiflexors weakness in normal subjects. *J. Electromyogr. Kinesiol.* **2009**, *19*, 774–781. [CrossRef]
21. Faul, F.; Erdfelder, E.; Lang, A.G.; Buchner, A. G*Power 3: A flexible statistical power analysis program for the social, behavioral, and biomedical sciences. *Behav. Res. Methods* **2007**, *39*, 175–191. [CrossRef]
22. Cohen, J. A power primer. *Psychol. Bull.* **1992**, *112*, 155–159. [CrossRef] [PubMed]
23. Pereira, R.; Machado, M.; dos Santos, M.M.; Pereira, L.N.; Sampaio-Jorge, F. Muscle activation sequence compromises vertical jump performance. *Serb. J. Sports Sci.* **2008**, *2*, 85–90.
24. Ribeiro, J.L.; de Castro, B.O.; Rosa, C.S.; Baptista, R.R.; Oliveira, A.R. Heart rate and blood lactate responses to changquan and daoshu forms of modern wushu. *J. Sports Sci. Med.* **2006**, *5*, 1–4. [PubMed]
25. Koshida, S.; Matsuda, T.; Kawada, K. Lower extremity biomechanics during kendo strike-thrust motion in healthy kendo athletes. *J. Sports Med. Phys. Fitness* **2011**, *51*, 357–365.
26. García-Manso, J.M.; Rodríguez-Matoso, D.; Sarmiento, S.; de Saa, Y.; Vaamonde, D.; Rodríguez-Ruiz, D.; Da Silva-Grigolettoc, M.E. Effect of high-load and high-volume resistance exercise on the tensiomyographic twitch response of biceps brachii. *J. Electromyogr. Kinesiol.* **2012**, *22*, 612–619. [CrossRef] [PubMed]
27. Park, S. Theory and usage of tensiomyography and the analysis method for the patient with low back pain. *J. Exerc. Rehabil.* **2020**, *16*, 325–331. [CrossRef] [PubMed]
28. Valencic, V.; Knez, N.; Simunic, B. Tensiomyography: Detection of skeletal muscle response by means of radial muscle belly displacement. *Biomed. Eng.* **2001**, *1*, 1–10.
29. Križaj, D.; Grabljevec, K.; Šimunič, B. Evaluation of muscle dynamic response measured before and after treatment of spastic muscle with a BTX-A: A case study. In Proceedings of the 11th Mediterranean Conference on Medical and Biological Engineering and Computing, Ljubljana, Slovenia, 26–30 June 2007; pp. 393–396. [CrossRef]
30. Pišot, R.; Narici, M.V.; Šimunič, B.; De Boer, M.; Seynnes, O.; Jurdana, M.; Mekjavić, I.B. Whole muscle contractile parameters and thickness loss during 35-day bed rest. *Eur. J. Appl. Physiol.* **2008**, *104*, 409–414. [CrossRef]
31. Chappell, J.D.; Herman, D.C.; Knight, B.S.; Kirkendall, D.T.; Garrett, W.E.; Yu, B. Effect of fatigue on knee kinetics and kinematics in stop-jump tasks. *Am. J. Sports Med.* **2005**, *33*, 1022–1029. [CrossRef]
32. Rusu, L.D.; Cosma, G.G.; Cernaianu, S.M.; Marin, M.N.; Rusu, P.F.; Ciocănescu, D.P.; Neferu, F.N. Tensiomyography method used for neuromuscular assessment of muscle training. *J. Neuroeng. Rehabil.* **2013**, *10*, 67. [CrossRef]
33. Valenčič, V.; Knez, N. Measuring of skeletal muscles' dynamic properties. *Artif. Organs.* **1997**, *21*, 240–242. [CrossRef] [PubMed]

 medicina

Article

Differences in Femoral Artery Occlusion Pressure between Sexes and Dominant and Non-Dominant Legs

Nicole D. Tafuna'i [1], Iain Hunter [1], Aaron W. Johnson [1], Gilbert W. Fellingham [2] and Pat R. Vehrs [1,*]

1 Department of Exercise Sciences, Brigham Young University, Provo, UT 84602, USA; nicoletafunai@gmail.com (N.D.T.); iain_hunter@byu.edu (I.H.); wayne_johnson@byu.edu (A.W.J.)
2 Department of Statistics, Brigham Young University, Provo, UT 84602, USA; gilbert_fellingham@byu.edu
* Correspondence: pat_vehrs@byu.edu

Abstract: *Background and Objectives*: Blood flow restriction during low-load exercise stimulates similar muscle adaptations to those normally observed with higher loads. Differences in the arterial occlusion pressure (AOP) between limbs and between sexes are unclear. We compared the AOP of the superficial femoral artery in the dominant and non-dominant legs, and the relationship between blood flow and occlusion pressure in 35 (16 males, 19 females) young adults. *Materials and Methods*: Using ultrasound, we measured the AOP of the superficial femoral artery in both legs. Blood flow at occlusion pressures ranging from 0% to 100% of the AOP was measured in the dominant leg. *Results*: There was a significant difference in the AOP between males and females in the dominant (230 ± 41 vs. 191 ± 27 mmHg; $p = 0.002$) and non-dominant (209 ± 37 vs. 178 ± 21 mmHg; $p = 0.004$) legs, and between the dominant and non-dominant legs in males (230 ± 41 vs. 209 ± 37 mmHg; $p = 0.009$) but not females (191 ± 27 vs. 178 ± 21 mmHg; $p = 0.053$), respectively. Leg circumference was the most influential independent predictor of the AOP. There was a linear relationship between blood flow (expressed as a percentage of unoccluded blood flow) and occlusion pressure (expressed as a percentage of AOP). *Conclusions*: Arterial occlusion pressure is not always greater in the dominant leg or the larger leg. Practitioners should measure AOP in both limbs to determine if occlusion pressures used during exercise should be limb specific. Occlusion pressures used during blood flow restriction exercise should be chosen carefully.

Keywords: resistance exercise; blood flow restriction; blood flow restriction exercise

Citation: Tafuna'i, N.D.; Hunter, I.; Johnson, A.W.; Fellingham, G.W.; Vehrs, P.R. Differences in Femoral Artery Occlusion Pressure between Sexes and Dominant and Non-Dominant Legs. *Medicina* **2021**, 57, 863. https://doi.org/10.3390/medicina57090863

Academic Editor: Filipe Manuel Clemente

Received: 23 July 2021
Accepted: 21 August 2021
Published: 24 August 2021

Publisher's Note: MDPI stays neutral with regard to jurisdictional claims in published maps and institutional affiliations.

1. Introduction

Blood flow restriction (BFR) applied to the arms or legs during low-load resistance training is effective in promoting hypertrophy and increasing or maintaining muscle strength [1–5]. Blood flow restriction exercise (BFRE) can be part of musculoskeletal rehabilitation following an injury or surgery or for those trying to counter muscle wasting due to chronic disease [4,6,7]. The muscular adaptations to BFRE contribute to the popularity of this method of resistance training among athletes and in the fitness industry.

Blood flow restriction partially restricts arterial blood flow into the limb and occludes venous blood flow out of the muscle [6–9]. Some studies [10,11] used elastic wraps as a "practical" method of BFR, but this could produce inconsistent blood flow restriction between two limbs and between exercise sessions. Most studies have restricted blood flow with an inflatable cuff. Early studies used absolute cuff pressures [7,8,12–16] ranging from 50 to 300 mmHg, but this is problematic in that a given cuff pressure represents a different level of occlusion and blood flow restriction for each person. The current recommendation [5,17] is to use a percentage of the arterial occlusion pressure (AOP) to restrict blood flow during BFRE. Although further research is needed to determine the optimal pressure to use during BFRE, it appears that a pressure equivalent to 50% to 80% of the AOP is appropriate during low-load resistance training [5].

Although a plethora of papers have been published on the topic of BFR and BFRE, some things remain unclear, including the relationship between blood flow and occlusion pressure, sex differences in AOP, and differences in AOP between dominant and non-dominant limbs. Differences in the AOP between individuals is attributed primarily to differences in limb circumference [5,18]. To date, studies have not reported the AOP in the dominant and non-dominant limbs in an individual. In most of the literature where occlusion pressure is based on a percentage of AOP, dominance of the occluded limb is not reported in unilateral interventions and any differences between limbs are not reported in bilateral studies [19–24]. Any differences in AOP between the dominant and non-dominant limbs may be due to differences in limb circumference. Although previous studies have included male and female participants, sex differences in AOP have not been reported. After accounting for potential differences in limb circumference, little evidence suggests that there is a sex difference in AOP. Some authors have reported that the relationship between arterial blood flow and absolute cuff pressure in the leg is linear [7,8,12,25] but a recent study [26] reported a nonlinear relationship between relative arterial blood flow and cuff pressures with a plateau in blood flow at pressures between 40–80% AOP.

The purposes of this study were to compare the AOP of the superficial femoral artery (SFA) in the dominant and non-dominant legs, and blood flow at relative occlusion pressures (0–100% AOP) in the dominant leg at rest in young healthy men and women. We hypothesized a direct positive relationship between limb circumference and AOP, no significant difference in AOP between the sexes or between the dominant and non-dominant legs, and no significant sex difference in the relationship between arterial blood flow and relative occlusion pressure.

2. Materials and Methods

This study was a cross-sectional study that measured the AOP in the SFA of both legs and blood flow in the dominant leg at cuff pressures representing 0%, 20%, 40%, 60%, 80%, and 100% of the AOP. The primary variables of interest included the AOP (mmHg) and blood flow at each increment of blood flow restriction (% AOP). This study was reviewed and approved by the Institutional Review Board for the use of Human Subjects prior to the collection of any data.

2.1. Participants

A total of 35 (16 males, 19 females) physically active and apparently healthy adults, 18–35 years of age participated in this study. Interested participants were excluded from participation if they had any known risk factors for cardiovascular disease or one or more risk factors for thromboembolism, which include: obesity (BMI $\geq$ 30 kg/m^2), diagnosed Crohn's disease, a previous fracture of the hip, pelvis, or femur, a major surgery in the last 6 months, varicose veins, a family history of deep vein thrombosis or pulmonary embolism, and on oral birth control [22,27–29]. Individuals were also excluded if a) they had been diagnosed as having or were being treated for cardiovascular disease, renal disease, diabetes, or hypertension, b) their resting systolic blood pressure (SBP) $\geq$ 130 or diastolic blood pressure (DBP) $\geq$ 80 mmHg, or c) they were pregnant or less than 6 months postpartum. To minimize the effects of hormone variability later in the menstrual cycle, females participated in the study within the first 14 days of their menstrual cycle.

2.2. Procedures

Subjects were instructed to refrain from eating during the 2 h prior to their participation, consuming caffeine for the previous 8 h, and participating in vigorous physical activity the previous 24 h [28,29]. All procedures for each subject were completed in one visit to the lab. The methods, expectations, risks, and benefits of the study were explained to each subject after which they voluntarily provided written informed consent.

The subject's height (cm) was measured using a calibrated wall-mounted stadiometer scale (SECA Model 264; SECA, Cino, CA, USA). Body mass (kg) was measured using a digital scale (Ohaus Model CD-33, Ohaus Corporation, Pine Brook, NJ, USA) and BMI (kg/m^2) was calculated from measured height and body mass values. Subjects then sat quietly in a comfortable chair for 5 min with legs uncrossed. Blood pressure was measured on the right arm and the average of two blood pressure measurements was recorded, or if they were not within 5 mmHg of each other, blood pressure was measured a third time, and the two closest measurements were averaged. Mean arterial pressure (MAP) was calculated as DBP plus one-third of pulse pressure. Leg dominance was determined by self-report by asking "I you were to kick a ball, with which leg would you use to kick the ball? [30]. The circumference and skinfold thickness of the dominant and non-dominant thighs were measured in triplicate in the standing position using a spring-loaded Gullick measuring tape and a calibrated Lange caliper (Santa Cruz, CA, USA), respectively. Measurements were taken at one-third of the distance between the inguinal crease and the top of the patella. The average of the three measurements was used in the data analysis.

2.3. *Blood Flow Measurements*

All blood flow measurements were performed using a handheld Doppler probe (9 MHz; 55 mm) and GE ultrasound machine with an integrated ECG (GE LOGIQ, GE Healthcare). Blood flow restriction was accomplished using a Hokanson SC10 cuff (10 cm wide; 85 cm long) attached to an E-20 rapid cuff inflator (Hokanson, Bellevue, WA, USA). The occlusion cuff was placed on the participant's thigh one-third of the distance between the inguinal fold and the top of the patella and blood flow in the SFA was measured distal to the cuff. Color flow mode and pulse wave forms were viewed to determine the presence of blood flow. During the entire time of testing, participants were in a semi-reclined (15°) position to allow reasonable access to the SFA using the ultrasound. Angle of insonation of the ultrasound probe was maintained at 60°.

2.4. *Measurement of Arterial Occlusion Pressure*

The AOP of the SFA in the dominant and non-dominant legs was measured once in a randomized order for each participant. A hand-held Doppler probe was used to detect a pulse wave in the SFA distal to the cuff with the cuff deflated. The cuff was then inflated to 50 mmHg and then gradually increased until arterial flow and pulse waves were no longer detected. After the AOP was recorded, the cuff was deflated, removed, and placed on the other leg. The participant rested for 5 min [20,27,29] with the cuff deflated, after which the process was repeated.

2.5. *Measurement of Arterial Blood Flow*

Following at least 5 min after the second AOP measurement, we measured arterial blood flow for 1 min at cuff pressures equivalent to 0%, 20%, 40%, 60%, 80%, and 100% of the subject's previously measured AOP in a randomized order. There was a 5 min rest period between measurements with the cuff deflated. One-minute video clips were stored for later analysis. Using the integrated ECG and pulse waves as reference points, femoral artery diameter was measured at two time periods representing the end of diastole (just before the QRS) and during systole (at the peak of the QRS) of each cardiac cycle. The two measurements were averaged for each beat over five 12-s periods. Time averaged blood flow velocity (TAV) over the five 12-s periods was recorded. Blood flow (mL/min) was calculated automatically by the ultrasound machine as follows:

$$\text{Blood flow (mL/min)} = \text{Cross sectional area (cm}^2) \times \text{TAV (cm/s)} \times 60\ \text{s/min}$$

2.6. *Data Analysis*

Sex differences in age, height, body mass, BMI, blood pressure measurements (i.e., SBP, DBP, MAP), leg circumference, thigh skinfold thickness, and AOP in the dominant and non-dominant legs were determined using two-sample *t*-tests. Differences in leg circumference, thigh skinfold thickness, and AOP between the dominant and non-dominant legs in males and females were determined using paired *t*-tests. The influence of sex, SBP, DBP,

MAP, thigh skinfold and circumference measurements on the AOP was evaluated using regression analysis.

Analysis of arterial blood flow data, expressed as a percentage of unoccluded blood flow, and occlusion pressure, expressed as a percentage of individual AOP (0%, 20%, 40%, 60%, 80%, 100%) presented two major challenges. The first was that relative blood flow when there was no occlusion (0% AOP) is represented as 100% for every subject and there is no variance in the data. Second, blood flow at various degrees of occlusion (e.g., 20%, 40%, 60% AOP) for some subjects was higher than when there was no occlusion (0% AOP). Thus, the difficulty in analyzing the blood flow data was that there was one data point (0% AOP) where there is no variance in blood flow (blood flow = 100%) and other data points where relative blood flow was greater than that measured at 0% AOP (blood flow = 100%). To analyze these data, we first used a one-sample *t*-test to determine if relative blood flow at 20% AOP was significantly different from relative blood flow at 0% AOP. We found that the average relative blood flow at 20% AOP was 81% (CI = 70.4−91.6%) of unoccluded blood flow ($p = 0.0009$). Since each subject had multiple data points, we then fit a mixed linear model between relative blood flow and relative occlusion pressure to account for within- and between-subject variability. To further appropriately account for variability when fitting the model, we omitted the blood flow data at 0% AOP and only used data at occlusion pressures of 20%, 40%, 60%, 80%, and 100% of AOP. The initial analysis revealed that there was no sex difference in the relationship between blood flow and occlusion pressure. We therefore fit a linear model that did not include sex as a variable. A 95% confidence interval (CI) and prediction interval (PI) were computed for the line of best fit through the data.

3. Results

Participant characteristics are shown in Table 1. Males were taller, heavier and had higher SBP and MAP than their female counterparts. There were no significant differences in the circumferences of the dominant and non-dominant legs in males ($p = 0.1897$) or in females ($p = 0.0895$) or of the dominant ($p = 0.847$) or non-dominant legs ($p = 0.746$) between males and females. There were no significant differences in the thigh skinfold between the dominant and non-dominant legs of either males ($p = 0.7630$) or females ($p = 0.5923$) or of the dominant leg ($p = 0.056$) and non-dominant leg ($p = 0.054$) between males and females.

Table 1. Participant Characteristics.

	Males (*N* = 16)	Females (*N* = 19)	Combined (*N* = 35)
Age (years)	23.8 ± 3.6	22.9 ± 3.5	23.3 ± 3.5
Height (cm) *	177.6 ± 5.3	166.9 ± 8.7	171.8 ± 9.1
Body Mass (kg) *	75.4 ± 10.6	63.6 ± 9.6	68.8 ± 11.6
BMI (kg/m^2)	23.9 ± 3.6	22.7 ± 2.9	23.3 ± 3.2
SBP (mmHg) *	122 ± 5.5	114 ± 5	117 ± 6.5
DBP (mmHg)	73 ± 6.4	71 ± 6	72 ± 6.3
MAP (mmHg) *	90 ± 5.7	85 ± 5	87 ± 5.6
Thigh Skinfold (mm)			
Dominant Leg	25.8 ± 11.0	32.6 ± 9.2	29.5 ± 10.5
Non-dominant Leg	25.9 ± 11.6	32.8 ± 8.8	29.6 ± 10.6
Difference	0.12 ± 1.63	0.26 ± 2.1	0.2 ± 1.87
Thigh Circumference (cm)			
Dominant Leg	48.8 ± 4.2	49.2 ± 4.9	49.0 ± 4.5
Non-dominant Leg	47.9 ± 4.5	48.5 ± 4.9	48.3 ± 4.7
Difference	0.91 ± 2.6	0.68 ± 1.6	0.78 ± 2.12

Table 1. *Cont.*

	Males (N = 16)	Females (N = 19)	Combined (N = 35)
Arterial Occlusion Pressure (AOP)			
Dominant Leg *	230 ± 41	191 ± 27	209 ± 39
Non-dominant Leg *	209 ± 37	178 ± 21	192 ± 33
Difference	21 ± 28.7 ^	13 ± 27.3	17 ± 27.8

Values are mean ± SD. BMI = body mass index, SBP = systolic blood pressure, DBP = diastolic blood pressure, MAP = mean arterial pressure. * = significant difference between males and females ($p < 0.05$); ^ = significant difference between dominant and non-dominant leg ($p < 0.05$).

3.1. Arterial Occlusion Pressure

There was a significant difference in the AOP between males and females in the dominant (230 ± 41 vs. 191 ± 27 mmHg; $p = 0.002$) and non-dominant (209 ± 37 vs. 178 ± 21 mmHg; $p = 0.004$) legs, respectively. There was a significant difference in the AOP between the dominant and non-dominant legs in males (230 ± 41 vs. 209 ± 37 mmHg; $p = 0.009$) but not in females (191 ± 27 vs. 178 ± 21 mmHg; $p = 0.053$). Regression analysis revealed that after leg circumference entered the equation, SBP, DBP, MAP, skinfold thickness, age, and sex were not significant independent predictors of AOP. The resulting regression model as shown in Figure 1 with 95% CI and 95% PI was:

$$\text{AOP (mmHg)} = 40.4 + (3.23)\ \text{Leg Circumference (cm)}$$

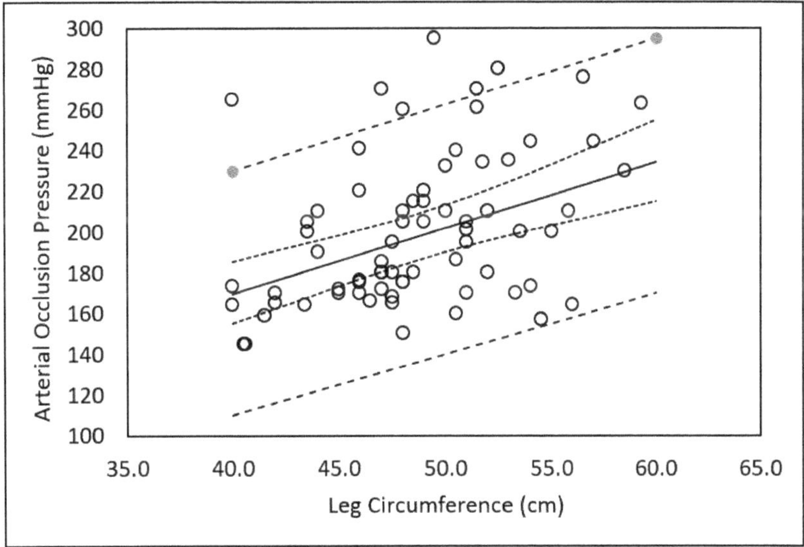

Figure 1. Relationship between limb circumference and arterial occlusion pressure. Solid line = line of best fit. Dashed lines = 95% Prediction Intervals. Dotted lines = 95% Confidence Intervals.

3.2. Arterial Blood Flow

The mixed model analysis revealed a linear relationship between relative blood flow (% unoccluded blood flow) and relative occlusion pressure (%AOP). The resulting equation (R = −0.842; Residual Standard Error = 25.3) as shown in Figure 2 with 95% CI and 95% PI was:

$$\text{Percent Blood Flow} = 99.46 - 0.85\ (\text{Occlusion Pressure; \%AOP})$$

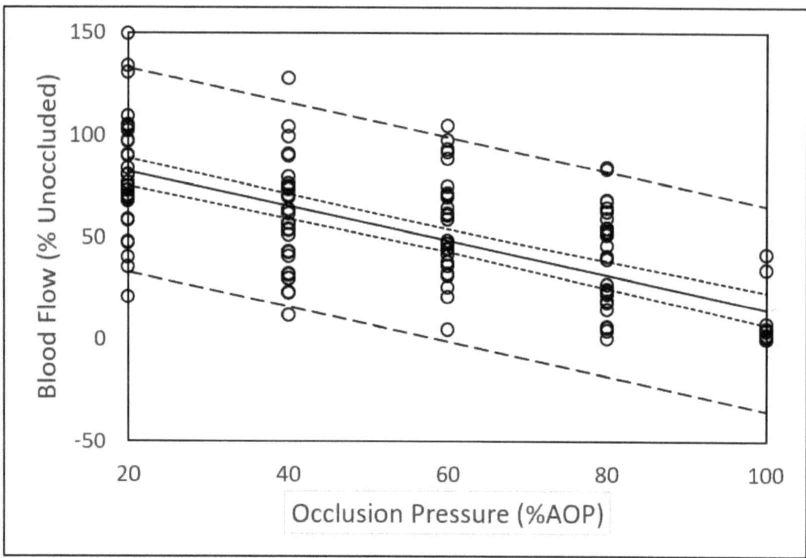

Figure 2. Relationship between arterial blood flow and occlusion pressure. Solid line = line of best fit. Dashed lines = 95% Prediction Intervals. Dotted lines = 95% Confidence Intervals.

4. Discussion

This paper adds to the current body of knowledge about BFR in that we report, perhaps for the first time, large differences in AOP between males and females and between the dominant and non-dominant legs. The linear relationship between blood flow and occlusion pressure expressed in relative terms was unrelated to sex. We also report a large variance in blood flow data at different levels of occlusion that is not unique to this study but has not been previously discussed. The findings of this study have implications for future research and those using BFRE.

4.1. Sex and Limb Differences in Arterial Occlusion Pressure

We report a large sex difference in AOP in both the dominant and non-dominant legs (Table 1). Although other studies have included male and female participants, sex differences in AOP have not been reported [21,26,31,32]. To the best of our knowledge, only one previous study has reported a sex difference in AOP. Jessee et al. [27] reported that the AOP of the right arm of females was on the average 4–7 mmHg ($p < 0.05$) lower than in males across three different cuff sizes. Although significantly different, the authors suggest that the differences in AOP were inconsequential in prescribing BFRE.

It is well reported that differences in the AOP can be attributed primarily to differences in limb circumference [7,18,33]. The larger the limb, the greater the pressure required to occlude the blood vessel. Larger limbs have more mass between the skin and the blood vessels that must be compressed to occlude the vessel, and higher pressures are required to transmit adequate force to the deeper tissues [20,21]. Hence, it follows that sex differences in AOP or differences in AOP between limbs may be accounted for by differences in limb circumference. Jessee et al. [27] reported an average sex difference in circumference of the right arm of 5.3 cm and that after accounting for arm circumference, arm length, SBP, and DBP, sex remained a significant independent predictor of AOP. In this study, there was an average difference in circumference of <1 cm in both the dominant and non-dominant legs between and within males and females (Table 1). Despite a small average difference in leg circumference, there was a large difference in AOP between the dominant and non-dominant legs within and between males and females (Table 1). This could be attributed to the fact that the difference in the circumferences of the dominant and non-dominant leg ranged from the dominant leg being 5.5 cm smaller to 7 cm larger than the non-dominant leg in males and 2.5 cm smaller to 3.5 cm larger in females. The regression analysis in this study indicates that after accounting for leg circumference, SBP, DBP, MAP, skinfold thickness, sex, and age were not significant predictors of AOP. The differences in the results between this study and that of Jessee et al. [27] might be explained by the differences in the size of the limbs studied (i.e., legs vs. arms). In addition, we report the AOP of both the dominant and non-dominant legs, whereas Jessee et al. only reported the AOP of the right arm, rather than the dominant arm. It should be appreciated that the leg circumference of the dominant leg is not always larger than that of the non-dominant leg. In this study, the dominant leg was larger than the non-dominant leg in 20 of the participants (8 males, 12 females) and the non-dominant leg was larger in 15 of the participants (8 males, 7 females). Likewise, the AOP is not always higher in the larger leg. In this study, the AOP was higher in the larger leg of 21 participants (15 dominant, 6 non-dominant; 9 males, 12 females) and higher in the smaller leg of 14 participants (9 dominant, 5 non-dominant; 7 males, 7 females).

Considering that the overall average difference in AOP between the dominant and non-dominant legs in this study was 17 mmHg (Table 1), an occlusion pressure of 50% of AOP would result in a difference in occlusion pressure of less than 9 mmHg between legs. This small difference in occlusion pressure between the two limbs would be of little import during BFRE. Nevertheless, the greatest difference in AOP between the two legs in an individual in this study was 80 mmHg. In this subject, the difference in occlusion pressure between the two legs during BFRE would be of practical significance. Our data suggest that AOP should be measured in both legs to determine if a sufficient difference existed that would justify using occlusion pressures specific to each leg. Most practitioners (e.g., physical therapists, personal trainers, strength and conditioning coaches, etc.) are unable to measure AOP in their clients or patients. Nevertheless, health and fitness professionals should be aware of differences between limbs that could affect the safe use of BFR during exercise. End-users should not be naïve of the potential differences between limbs and should use BFR during exercise with appropriate caution. Although further research is needed, occlusion pressures at a perceived pressure [10,11] could account for difference between limbs when measures of occlusion pressure are not possible.

The composition of the limb may also influence the pressure required to occlude a blood vessel. In this study, the sex differences in thigh skinfold thickness approached the alpha-level of 0.05 but likely did not enter into the regression equation to estimate AOP because it is included as part of the overall circumference of the leg and represents only a portion of the total tissue mass that must be compressed to occlude the femoral artery. Our data concur with that of Loenneke et al. [21] who, after using B-mode ultrasound to measure fat thickness of the upper arm of 171 males and females, concluded that the absolute size of the arm may be more important than the composition of the arm in predicting AOP.

4.2. *Arterial Blood Flow*

Since blood flow at any given absolute pressure varies widely between individuals, it is appropriate to express blood flow and occlusion pressure in relative terms. The results of this study (Figure 2) indicate a linear relationship between blood flow (% unoccluded blood flow) and relative occlusion pressure (%AOP). Our data concur with those of a previous study reporting a linear relationship between relative blood flow and %AOP in the posterior tibial artery [25]. This is contrary to recently reported nonlinear relationships between relative blood flow and relative occlusion pressure and plateaus in blood flow between 40% to 80% AOP in the brachial artery [34,35] and the SFA [26]. Some difference in methodology between studies could help explain the disparity in the results. For example, subjects in our study were in a semi reclined position, whereas subjects in Crossley et al. [26] study were in the seated position. Additionally, measurements in the study by Crossley et al. were performed on alternating legs in a randomized order over the course of the study so the reported AOP and blood flow represented AOP and blood flow in both legs rather than either the dominant or non-dominant leg.

Whether the relationship between blood flow and occlusion is linear or nonlinear is of practical importance. A nonlinear relationship suggests that use of a lower, more comfortable and potentially safer occlusion pressure (e.g., 40% AOP) would provide an equally effective stimulus during BFRE as higher occlusion pressures. A linear relationship suggests that the occlusion pressure used during BFRE should be selected more carefully and that further research is required to determine a recommended reduction in blood flow to be used during BFRE.

4.3. *Variance in Blood Flow Measurements*

In this study, we observed a large variation in blood flow at different levels of occlusion. For example, we note that blood flow at higher levels of occlusion was sometimes greater than at lower levels of occlusion. We also found that some participants had notable blood flow at an occlusion pressure equivalent to the previously measured AOP. Variance in the data presented in this study is apparent in the wide prediction intervals shown in Figure 2. These observations are suggestive of a robust cardiovascular system that maintains blood flow across various levels of occlusion pressures [35].

Evidence of the variation in blood flow measurements is present in previous studies. For example, close examination of previously reported blood flow data [34] reveals that relative blood flow at 70% AOP was greater than relative blood flow at 60% and 50% AOP. Likewise, previously reported large standard deviations of blood flow data [35] suggest that in some subjects, blood flow at higher occlusion pressures was greater than at lower occlusion pressures. This could be attributed to a cardiovascular response to high occlusion pressures in the absence of exercise [7,36]. It could also be possible that after several applications of BFR there are local responses in the vasculature that alters blood flow or the AOP. Although previous research indicates that blood flow returns to normal within 30 to 90 s after the occlusion is removed [37], longer rest periods may be needed between sequential blood flow measurements with occlusion. It is possible that after multiple occlusions of blood flow, the AOP changes. This could influence the expression of blood flow relative to AOP. Lastly, although data collected from each subject in this study occurred in a single day, Mouser et al. [35] reported a significant day-to-day variation in resting blood flow that clearly has implications for future research involving blood flow measurements over multiple days and the use of BFRE. Our data and close examination of data presented in the literature warrants a call for further studies evaluating the variance and reliability of blood flow measurements during BFR.

4.4. *Study Limitations*

This study had several limitations. Participants were college-age coeds without known risk factors for cardiometabolic diseases. Therefore, the results of our study may not be applicable to all populations. Blood pressure was not measured during the measurement of blood flow at different occlusion pressures. Having blood pressure measurements could lead to a better understanding of the relationship between occlusion pressure and blood flow. Blood pressure measurements during blood flow occlusion may also help explain the variation in blood flow at different occlusion pressures. In this study, the Hokanson SC10 cuff (10 cm wide; 85 cm long) attached to an E-20 rapid cuff inflator (Hokanson, Bellevue, WA, USA) was used for all measurements. Clinicians, researchers, and other practitioners may use different brands of cuffs and inflation systems, different cuff sizes, or other methods to occlude blood flow. Lastly, a greater number of subjects could improve the data when comparing limb and sex differences in AOP.

4.5. *Direction for Future Studies*

Based on the results of this study, future studies should include both male and female participants and report limb dominance, AOP, and blood flow data on both limbs in males and females. Limb circumference or other measures of limb volume should also be reported. While it is clear that limb circumference is more influential on AOP than limb composition, the sex differences in thigh skinfold thickness reported in this study and the influence of fat thickness reported by Loenneke et al. [21] lend support for an influence of limb composition on AOP that needs further investigation. The large variation of blood flow measurements at different occlusion pressures reported in this and previous studies suggests the need to standardize blood flow measurement methods and investigate the reliability of AOP and blood flow measurements. Comparing blood flow and variability in blood flow at different occlusion pressures between the dominant and non-dominant leg is also warranted. Measuring blood pressure during occlusion may help explain variation in blood flow at different occlusion pressures. Assessments of reliability in measurements of blood flow with and without occlusion between test administrators, within and between days needs attention.

5. Conclusions

An important finding of this study was large sex differences in AOP in both the dominant and non-dominant legs and large differences in AOP between the dominant and non-dominant legs particularly in men. Arterial occlusion pressure is not always greater in the dominant leg or the larger leg. Practitioners should measure AOP in both limbs to determine if occlusion pressures used during exercise should be limb specific. We also report a linear relationship between relative occlusion pressure and blood flow. The large variance in blood flow at different occlusion pressured warrants further study and caution during BFRE. These findings are of practical importance when using BFRE in various settings and suggest the need for continued research. Occlusion pressures used during blood flow restriction exercise should be chosen carefully.

Author Contributions: Conceptualization, N.D.T., I.H., P.R.V. and A.W.J.; methodology, N.D.T., A.W.J., G.W.F. and P.R.V.; formal analysis, N.D.T., A.W.J., G.W.F., P.R.V.; investigation, N.D.T., A.W.J. and P.R.V.; resources, P.R.V. and A.W.J.; writing—original draft preparation, N.D.T. and P.R.V.; writing—review and editing, I.H., A.W.J. and G.W.F.; project administration, N.D.T. and P.R.V. All authors have read and agreed to the published version of the manuscript.

Funding: This research received no external funding.

Institutional Review Board Statement: The study was conducted according to the guidelines of the Declaration of Helsinki and approved by the Institutional Review Board of Brigham Young University (Protocol X19019; 5 March 2019).

Informed Consent Statement: Informed written consent was obtained from all subjects involved in the study.

Data Availability Statement: The data presented in this study are available upon request from the corresponding author. The data are not publicly available.

Conflicts of Interest: The authors declare no conflict of interest.

References

1. Loenneke, J.P.; Wilson, J.M.; Marin, P.J.; Zourdos, M.C.; Bemben, M.G. Low intensity blood flow restriction training: A meta-analysis. *Eur. J. Appl. Physiol.* **2012**, *112*, 1849–1859. [CrossRef] [PubMed]
2. Staunton, C.A.; May, A.K.; Brandner, C.R.; Warmington, S.A. Haemodynamics of aerobic and resistance blood flow restriction exercise in young and older adults. *Eur. J. Appl. Physiol.* **2015**, *115*, 2293–2302. [CrossRef] [PubMed]
3. Loenneke, J.P.; Fahs, C.A.; Wilson, J.M.; Bemben, M.G. Blood flow restriction: The metabolite/volume threshold theory. *Med. Hypotheses* **2011**, *77*, 748–752. [CrossRef] [PubMed]
4. Kim, D.; Loenneke, J.P.; Ye, X.; Bemben, D.A.; Beck, T.W.; Larson, R.D.; Bemben, M.G. Low-load resistance training with low relative pressure produces muscular changes similar to high-load resistance training. *Muscle Nerve* **2017**, *56*, E126–E133. [CrossRef] [PubMed]
5. Scott, B.R.; Loenneke, J.P.; Slattery, K.M.; Dascombe, B.J. Exercise with blood flow restriction: An updated evidence-based approach for enhanced muscular development. *Sports Med.* **2015**, *45*, 313–325. [CrossRef]

6. Pope, Z.K.; Willardson, J.M.; Schoenfeld, B.J. Exercise and blood flow restriction. *J. Strength Cond. Res.* **2013**, *27*, 2914–2926. [CrossRef] [PubMed]
7. Iida, H.; Takano, H.; Meguro, K.; Asada, K.; Oonuma, H.; Morita, T.; Kurano, M.; Sakagami, F.; Uno, K.; Hirose, K.; et al. Hemodynamic and autonomic nervous responses to the restriction of femoral blood flow by KAATSU. *Int. J. KAATSU Train. Res.* **2005**, *1*, 57–64. [CrossRef]
8. Iida, H.; Kurano, M.; Takano, H.; Kubota, N.; Morita, T.; Meguro, K.; Sato, Y.; Abe, T.; Yamazaki, Y.; Uno, K.; et al. Hemodynamic and neurohumoral responses to the restriction of femoral blood flow by KAATSU in healthy subjects. *Eur. J. Appl. Physiol.* **2007**, *100*, 275–285. [CrossRef]
9. Loenneke, J.P.; Pujol, T.J. The Use of Occlusion Training to Produce Muscle Hypertrophy. *Strength Cond. J.* **2009**, *31*, 77–84. [CrossRef]
10. Lowery, R.P.; Joy, J.M.; Loenneke, J.P.; de Souza, E.; Machado, M.; Dudeck, J.E.; Wilson, J.M. Practical blood flow restriction training increases muscle hypertrophy during a periodized resistance training programme. *Clin. Phsiol. Funct. Imaging* **2014**, *34*, 317–321. [CrossRef] [PubMed]
11. Wilson, J.M.; Lowery, R.P.; Joy, J.M.; Loenneke, J.P.; Naimo, M.A. Practical blood flow restriction training increases acute determinants on hypertrophy without increasing indices of muscle damage. *J. Strength Cond. Res.* **2013**, *27*, 3068–3075. [CrossRef] [PubMed]
12. Hunt, J.E.; Stodart, C.; Ferguson, R.A. The influence of participant characteristics on the relationship between cuff pressure and level of blood flow restriction. *Eur. J. Appl. Physiol.* **2016**, *116*, 1421–1432. [CrossRef]
13. Meister, C.B.; Kutianski, F.A.T.; Carstens, L.C.; Andrade, S.L.F.; Rodacki, A.L.F.; Martins de Souza, R. Effects of two programs of metabolic resistance training on strength and hypertrophy. *Fisioterpia Mov.* **2016**, *29*, 147–155. [CrossRef]
14. Pope, Z.K.; Willardson, J.M.; Schoenfeld, B.J.; Emmett, J.; Owen, J.D. Hypertrophic and strength response to exxcentric resistance training with blood flow restriction. *Int. J. Sports Sci.* **2015**, *10*, 919–931.
15. Sumide, T.; Sakuraba, K.; Sawaki, K.; Ohmura, H.; Tamura, Y. Effect of resistance exercise training combined with relatively low vascular occlusion. *J. Sci. Med. Sport* **2009**, *12*, 107–112. [CrossRef] [PubMed]
16. Wernbom, M.; Augustsson, J.; Thomee, R. Effects of vascular occlusion on muscular endurance in dynamic knee extension exercise at different submaximal loads. *J. Strength Cond. Res.* **2006**, *20*, 372–377. [CrossRef] [PubMed]
17. McEwen, J.A.; Owens, J.G.; Jeyasurya, J. Why is it crucial to use personalized occlusion pressures in blood flow restriction (BFR) rehabilitation? *J. Med. Biol. Eng.* **2019**, *39*, 173–177. [CrossRef]
18. Jessee, M.B.; Buckner, S.L.; Mouser, J.G.; Mattocks, K.T.; Loenneke, J.P. Letter to the editor: Applying the blood flow restriction pressure: The elephant in the room. *Am. J. Physiol. Heart Circ. Physiol.* **2016**, *310*, H132–H133. [CrossRef]
19. Loenneke, J.P.; Thiebaud, R.S.; Fahs, C.A.; Rossow, L.M.; Abe, T.; Bemben, M.G. Blood flow restriction does not result in prolonged decrements in torque. *Eur. J. Appl. Physiol.* **2013**, *113*, 923–931. [CrossRef] [PubMed]
20. Loenneke, J.P.; Fahs, C.A.; Rossow, L.M.; Sherk, V.D.; Thiebaud, R.S.; Abe, T.; Bemben, D.A.; Bemben, M.G. Effects of cuff width on arterial occlusion: Implications for blood flow restricted exercise. *Eur. J. Appl. Physiol.* **2012**, *112*, 2903–2912. [CrossRef] [PubMed]
21. Loenneke, J.P.; Allen, K.M.; Mouser, J.G.; Thiebaud, R.S.; Kim, D.; Abe, T.; Bemben, M.G. Blood flow restriction in the upper and lower limbs is predicted by limb circumference and systolic blood pressure. *Eur. J. Appl. Physiol.* **2015**, *115*, 397–405. [CrossRef]
22. Barnett, B.E.; Dankel, S.J.; Counts, B.R.; Nooe, A.L.; Abe, T.; Loenneke, J.P. Blood flow occlusion pressure at rest and immediately after a bout of low load exercise. *Clin. Physiol. Funct. Imaging* **2016**, *36*, 436–440. [CrossRef] [PubMed]
23. Counts, B.R.; Dankel, S.J.; Barnett, B.E.; Kim, D.; Mouser, J.G.; Allen, K.M.; Thiebaud, R.S.; Abe, T.; Bemben, M.G.; Loenneke, J.P. Influence of relative blood flow restriction pressure on muscle activation and muscle adaptation. *Muscle Nerve* **2016**, *53*, 438–445. [CrossRef]
24. Brandner, C.R.; Kidgell, D.J.; Warmington, S.A. Unilateral bicep curl hemodynamics: Low-pressure continuous vs high-pressure intermittent blood flow restriction. *Scand. J. Med. Sci. Sports* **2015**, *25*, 770–777. [CrossRef] [PubMed]
25. Mouser, J.G.; Dankiel, S.J.; Mattocks, K.T.; Jessee, M.B.; Buckner, S.L.; Abe, T.; Loenneke, J.P. Blood flow restriction and cuff width: Effect on blood flow in the legs. *Clin. Phsiol. Funct. Imaging* **2018**, *38*, 944–948. [CrossRef] [PubMed]
26. Crossley, K.W.; Porter, D.A.; Ellsworth, J.; Caldwell, T.; Feland, J.B.; Mitchell, U.H.; Johnson, A.W.; Egget, D.; Gifford, J.R. Effect of cuff pressure on blood flow during blood flow-restricted rest and exercise. *Med. Sci. Sports Exerc.* **2019**, *52*, 746–753. [CrossRef]
27. Jessee, M.B.; Buckner, S.L.; Dankel, S.J.; Counts, B.R.; Abe, T.; Loenneke, J.P. The Influence of Cuff Width, Sex, and Race on Arterial Occlusion: Implications for Blood Flow Restriction Research. *Sports Med.* **2016**, *46*, 913–921. [CrossRef]
28. Mattocks, K.T.; Jessee, M.B.; Counts, B.R.; Buckner, S.L.; Grant Mouser, J.; Dankel, S.J.; Laurentino, G.C.; Loenneke, J.P. The effects of upper body exercise across different levels of blood flow restriction on arterial occlusion pressure and perceptual responses. *Physiol. Behav.* **2017**, *171*, 181–186. [CrossRef] [PubMed]
29. Sieljacks, P.; Knudsen, L.; Wernbom, M.; Vissing, K. Body position influences arterial occlusion pressure: Implications for the standardization of pressure during blood flow restricted exercise. *Eur. J. Appl. Physiol.* **2018**, *118*, 303–312. [CrossRef] [PubMed]
30. Van Melick, N.; Meddeler, B.M.; Hoogeboom, T.J.; Nijhuis-van der Sanden, M.W.G.; van Cingel, R.E.H. How to determine leg dominance: The agreement between self-reported and observed performance in healthy adults. *PLoS ONE* **2017**, *12*, e0189876. [CrossRef] [PubMed]

31. Spitz, R.W.; Chatakondi, R.N.; Bell, Z.W.; Wong, V.; Dankel, S.J.; Abe, T.; Loenneke, J.P. The impact of cuff width and biological sex on cuff prefference and the percieve discomfort to blood-flow-restricted exercise. *Physiol. Meas.* **2019**, *40*, 055001. [CrossRef] [PubMed]
32. Wong, V.; Abe, T.; Chatakondi, R.N.; Bell, Z.W.; Spitz, R.W.; Dankel, S.J.; Loenneke, J.P. The influence of biological sex and cuff width on muscle swelling, echo intensity, and the fatigue response to blood flow restricted exercise. *J. Sports Sci.* **2019**, *37*, 1865–1873. [CrossRef]
33. Scott, B.R.; Loenneke, J.P.; Slattery, K.M.; Dascombe, B.J. Blood flow restricted exercise for athletes: A review of available evidence. *J. Sci. Med. Sport* **2016**, *19*, 360–367. [CrossRef] [PubMed]
34. Mouser, J.G.; Ade, C.J.; Black, C.D.; Bemben, D.A.; Bemben, M.G. Brachial blood flow under relative levels of blood flow restriction in decreased in a nonlinear fashion. *Clin. Physiol. Funct. Imaging* **2018**, *38*, 425–430. [CrossRef] [PubMed]
35. Mouser, J.G.; Dankel, S.J.; Jessee, M.B.; Mattocks, K.T.; Buckner, S.L.; Counts, B.R.; Loenneke, J.P. A tale of three cuffs: The hemodynamics of blood flow restriction. *Eur. J. Appl. Physiol.* **2017**, *117*, 1493–1499. [CrossRef] [PubMed]
36. Spranger, M.D.; Krishnan, A.C.; Levy, P.D.; O'Leary, D.S.; Smith, S.A. Blood flow restriction training and the exercise pressor reflex: A call for concern. *Am. J. Physiol. Heart Circ. Physiol.* **2015**, *309*, H1440–H1452. [CrossRef] [PubMed]
37. Lewis, P.; Psaila, J.V.; Davies, W.T.; McCarty, K.; Woodcock, J.P. Measurement of volume flow in the human common femoral artery using a duplex ultrasound system. *Ultrasound Med. Biol.* **1986**, *12*, 777–784. [CrossRef]

 medicina MDPI

Article

Anabolic–Androgenic Steroid Abuse among Gym Users, Eastern Province, Saudi Arabia

Walied Albaker [1], Ali Alkhars [1], Yasir Elamin [1], Noor Jatoi [1], Dhuha Boumarah [1] and Mohammed Al-Hariri [2,*]

1 Department of Internal Medicine, College of Medicine, Imam Abdulrahman Bin Faisal University, P.O. Box 2114, Dammam 31451, Saudi Arabia; wialbakr@iau.edu.sa (W.A.); aakhars@iau.edu.sa (A.A.); yaelamin@iau.edu.sa (Y.E.); najatoi@iau.edu.sa (N.J.); dohanahar@gmail.com (D.B.)

2 Department of Physiology, College of Medicine, Imam Abdulrahman Bin Faisal University, P.O. Box 2114, Dammam 31451, Saudi Arabia

* Correspondence: mtalhariri@iau.edu.sa; Tel.: +96-650-727-5028

Abstract: *Background and Objectives*: The main aim of the present study was to assess the use of androgenic–anabolic steroids (AAS) and to investigate its potentially unfavorable effects among gym members attending gym fitness facilities in Eastern Province, Saudi Arabia. *Materials and Methods*: A cross-sectional questionnaire-based study was carried out during the summer of 2017. Male gym users in the Eastern Province region of Saudi Arabia were the respondents. Information on socio-demographics, use of AAS, knowledge, and awareness about its side effects were collected using a self-administered questionnaire. *Results:* The prevalence of AAS consumption among trainees in Eastern Province was 21.3%. The percentage was highest among those 26–30 years of age (31.9%), followed by the 21–25 (27.4%) (p = 0.003) age group. Participants in the study were not aware of the potential adverse effects of AAS use. Adverse effects experienced by 77% of AAS users include psychiatric problems (47%), acne (32.7%), hair loss (14.2%), and sexual dysfunction (10.7%). Moreover, it appears that trainers and friends are major sources (75.20%) for obtaining AAS. *Conclusion:* AAS abuse is a real problem among gym members, along with a lack of knowledge regarding its adverse effects. Health education and awareness programs are needed not only for trainees, but also for trainers and gym owners as they are reportedly some of the primary sources of AAS.

Keywords: anabolic steroids; gym members; male; gym; abuse

Citation: Albaker, W.; Alkhars, A.; Elamin, Y.; Jatoi, N.; Boumarah, D.; Al-Hariri, M. Anabolic–Androgenic Steroid Abuse among Gym Users, Eastern Province, Saudi Arabia. *Medicina* **2021**, 57, 703. https://doi.org/10.3390/medicina57070703

Academic Editors: Filipe Manuel Clemente and Jose Antonio de Paz

Received: 2 June 2021
Accepted: 7 July 2021
Published: 10 July 2021

1. Introduction

Androgenic–anabolic steroids (AAS) are synthetically occurring products of the male sex hormone (Testosterone) [1]. They have two major effects which are anabolic and androgenic in nature [2]. The anabolic effect leads to decreased body fat and increases bone density and skeletal muscle mass, as well as stimulating erythropoiesis [3]. The androgenic effects are associated with the development of male sexual characteristics [4]. They have a significant effect on athletic performance [1].

A higher level of AAS in the body can lead to several psychological and physical complications. Examples of the physical problems that AAS can lead to are high blood pressure, atherosclerosis, myocardial infarction, cardiac hypertrophy, fluid retention, jaundice, acne, and hepatic tumors [1,5]. Psychiatric problems such as aggressiveness, euphoria, irritability, and mood disturbance can occur. Furthermore, AAS can indeed cause reduced sperm count, shrinking of the testicles, infertility, baldness, and the development of prostate and breast cancers [6].

Most countries permit the use of AAS to treat medical conditions by prescription [7]. Studies show an increasing rate of AAS use among athletes worldwide. However, the World Anti-Doping Agency lists AAS as prohibited substances [8].

Many bodybuilders turn to these medications to increase muscle size, strength, and overall efficiency with less effort over a shorter time rather than relying on physical exercise and a healthy diet alone [9].

Data show that AAS abuse is more prevalent in Brazil, Western countries, and the Middle East, and is less prevalent in Asia and Africa [10]. A few studies have evaluated the prevalence and awareness about the use of AAS among bodybuilders in Saudi Arabia. Previously reported data found that AAS ranked the highest among Saudi athletes who tested positive for prohibited agents [5]. Another study evaluating AAS abuse among bodybuilders in the southern province of Saudi Arabia (Jazan) revealed a lifetime prevalence of 31% [11].

The prevalence of AAS abuse was lower in the 316 male gym users in the northwestern region of Saudi Arabia (11.4%) [12]. Meanwhile, many studies have concluded that a lack of knowledge is the most likely cause regarding the use of AAS and its adverse effects on Saudi gym members. These studies also recommended a national awareness program in the central (Riyadh) [5], western (Jeddah) [12], and southern (Jazan) [11] regions of Saudi Arabia. Therefore, the present study sought to assess the prevalence of androgenic–anabolic steroid (AAS) use, and to investigate motivations for use and knowledge of its potentially unfavorable effects among gym center visitors in Eastern Province, Saudi Arabia.

2. Materials and Methods

A cross-sectional survey-based study was conducted on participants visiting gymnasiums in the Eastern Province (Al-Hufof, Ad-Dammam, Al-Khobar, Al-Qatif, Al-Jubail, and Saihat), Kingdom of Saudi Arabia, during the summer of 2017.

The calculation of the sample size was based on a prevalence of AAS of 50%, with a 95% confidence interval (CI). The inclusion criteria were the following: male, older than 18 years, and willingness to take part. We excluded repeated registration in multiple centers or incomplete responses.

The Ethics Committee of Imam Abdulrahman Bin Faisal University approved the research proposal and questionnaire (IRB-2018-01-174, approval date, 27 September 2018). All the participants agreed to participate and signed a consent form before entering into the study.

2.1. Questionnaire

A self-administered questionnaire was used for data collection. The questionnaire was designed based on previous similar studies reported in the literature on the same topic [12,13]. A pilot study was then carried out to determine the reliability and validity of the questionnaire. The feedback was analyzed and a finalized questionnaire was created accordingly. The final version of the questionnaire consisted of questions on (1) sociodemographic characteristics; (2) overall knowledge about the most commonly used and adverse effects of anabolic steroids; (3) prevalence of side effects experienced by users; and (4) the practices and patterns of AAS use. A single set of questionnaires was distributed (Arabic version).

2.2. Statistical Analysis

The statistical analysis was performed using SPSS software version 23.0 (SPSS Inc., Chicago, IL, USA). Descriptive statistics were used to explain the categorical and outcome items. A comparison between subgroups was made using Pearson's Chi-square test. A p-value less than 0.05 was used to indicate statistical significance.

3. Results

3.1. Sociodemographic Characteristics

Nearly 597 eligible gym members were invited to participate in the present study. A total of 541 participants answered the questionnaire and were included in the study according to the calculated sample size, with an overall response rate of 90.6%. Table 1 shows the characteristics of the study sample. Most of the users (67.1%) belong to the 21–25 age group. However, the present data showed that 72.3% of the studied gym members were currently employed and received higher education, i.e., bachelor's degree or higher (67.1%). Moreover, the majority of the study participants were single (61%), and half of them (50.8%) drew income less than 5000 Saudi Riyal (SR) (Table 1).

Table 1. Sociodemographic characteristics (n = 541).

Characteristic	*N*	%
Age Group		
Less than 18	11	2
18–20	53	9.8
21–25	184	67.1
26–30	161	29.8
31–35	66	12.2
35 and older	66	12.2
Occupation		
Employed	391	72.3
Unemployed	150	27.7
Level of Education		
Intermediate or Lower	23	4.3
High School	155	28.7
Higher Education	363	67.1
Social Status		
Single	330	61
Married	211	39
Monthly Income Saudi Riyal (SR)		
Less than 5000	275	50.8
5000–10,000	173	32
More than 10,000	83	15.3

3.2. Practices Associated with Androgenic–Anabolic Steroid Consumption

As shown in Figure 1, the most commonly used AAS in our study across all ages were Anavar (61.9%), Dianabol (46%), and Deca Durabolin (45.1%).

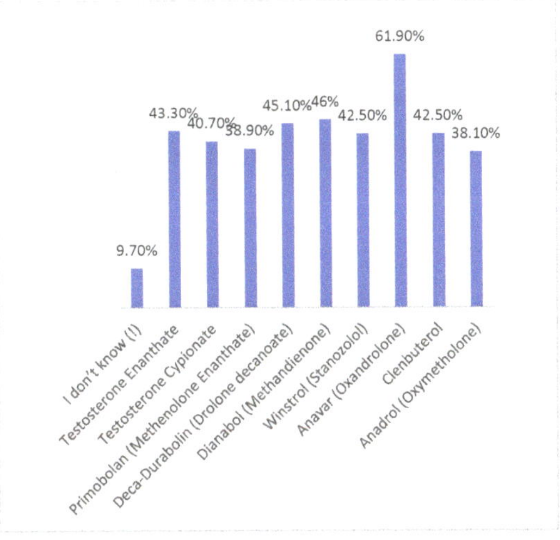

Figure 1. The prevalence of the most commonly used Androgenic–Anabolic Steroids (AAS) brands (generic).

Nearly 64.6% of gym members had been consuming AAS for more than 5 years, while 20.4% had been using steroids for three years or more (Figure 2).

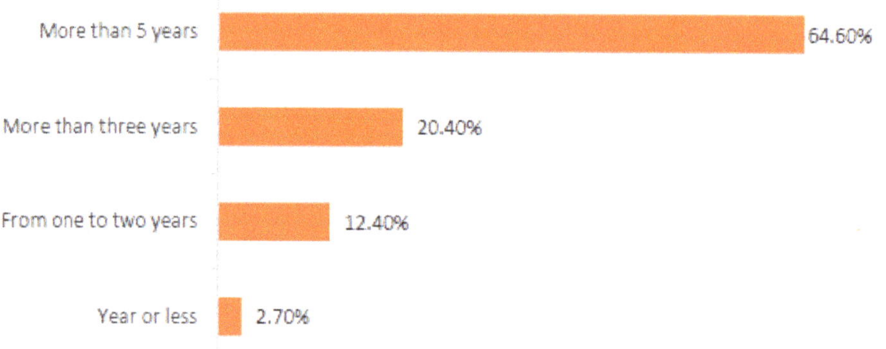

Figure 2. The relationship Between Using of Steroids and Years of Practice.

The majority of the AAS users (77%) reported side effects, and 47% experienced psychiatric problems, including depression, insomnia, and lower appetite. Acne was reported in 32.7% and hair loss in 14.2% of participants, as shown in Table 2.

Table 2. The prevalence of side effects experienced by users.

Side Effect	%
Any Side effect	77%
Acne	32.7%
Hair loss	14.2%
Breast Enlargement	9.7%
Psychiatric problems	47%
Sexual Dysfunction	10.6%
Chest pain	2.7%

The prevalence of AAS users in this study was 21.3%. Furthermore, it was higher in the age group of 26–30 (31.9%), followed by the 21–25 group (27.4%). The majority of them were employed (82.3%) with low (less than 5000 SR) income (41.6%), as presented in Table 3. According to Figure 3, it appears that trainers and friends are a major source for obtaining AAS.

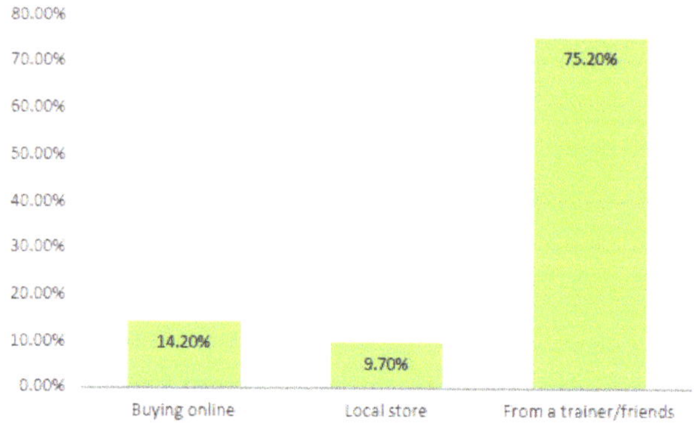

Figure 3. Sources of steroids.

Table 3. The prevalence of anabolic steroid use according to selected characteristics.

Characteristic	Users/Total	Prevalence%
Prevalence	113/541	21.3
Age Group		
Less than 18	0/11	0
18–20	5/53	4.4
21–25	31/184	27.4
26–30	36/161	31.9
31–35	19/66	16.8
35 and older	22/66	19.5
Occupation		
Employed	93/391	82.3
Unemployed	20/150	17.7
Level of Education		
Intermidate or lower	3/23	2.7
High School	35/155	31
Higher Education	75/363	66.4
Social Status		
Single	66/330	58.4
Married	47/211	41.6
Monthly Income		
Less than 5000	47/275	41.6
5000–10,000	43/173	38.1
More than 10,000	23/83	20.4

Approximately 63.7% of participating gym members observed an increase in activity, while 76.1% reported an increase in power after using AAS. However, 88.5% of users reported fast muscle mass gain (Figure 4).

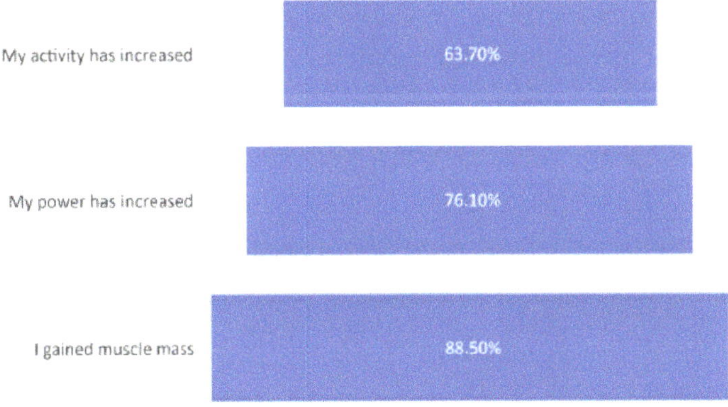

Figure 4. Reported advantages of using anabolic steroids.

3.3. Knowledge about and Attitude towards Androgenic–Anabolic Steroid Consumption

Surprisingly, the participants in our study were not aware of the potential adverse effects of AAS use. However, the majority of the AAS users had adequate knowledge of the adverse effects of AAS compared with nonusers. Approximately 38.1% of participants were aware that using AAS may lead to increased blood pressure, hair loss (34.5%), and acne (33.6%). Meanwhile, the nonusers expressed more knowledge of infertility (48.69%). About 18.6% of AAS users think that using AAS without a medical prescription is legal, while 15% did not know if it was legal to use ASS with a medical prescription or not. Among users, 66.4% believe it is illegal and still use it (Table 4).

Table 4. Knowledge about the adverse effects of anabolic steroids.

Parameter	Using Steroids			
	Users		Non-Users	
Acne	38	33.6%	86	20.1%
Heart Problems	19	16.8%	100	23.4%
Hair Loss	39	34.5%	86	20.1%
Increase Blood Pressure	43	38.1%	106	24.8%
Infertility	27	23.9%	208	48.6%
Liver Problems	29	25.7%	113	31.1%
There is No harm	7	6.2	0	0%
I don't know	24	21.2	160	37.4%
Legal	21	18.6%	29	6.8%
Illegal	75	66.4%	294	68.7%
I don't know	17	15%	105	24.5%

4. Discussion

In this study, we evaluated the prevalence, practices, knowledge, and attitudes of gym members in Eastern Province, Saudi Arabia, using AAS. The prevalence of AAS users among males attending gym centers was as high as 21.3%. These findings are in accordance with other studies conducted in the region (Kuwait and the UAE) [14,15]. Furthermore, two local studies were done in 2016 and 2017, in Riyadh [13] and Jazan [12], respectively. Both studies show an approximately 10% higher prevalence rate.

There are many reasons behind the prevalence of AAS consumption among gym members in the Eastern Province, such as competition among athletes to quickly build muscle mass. Respondents believed that the muscle growth advantages of AAS outweighed its adverse harmful effects [15], as the use of AAS makes lifting heavy objects easier [5]. Even some trainees recommended allowing the use of AAS for enhancing performance [16].

A local study found that the majority (77%) of athletes who self-declared AAS use and were aware of their adverse effects would still recommend them to friends [13]. Another reason for the high use rates could be due to the availability of AAS in gym centers and from trainers and friends. As reported recently, the prevalence of AAS consumption was highly influenced by its availability [16].

The highest prevalence of AAS usage in our study was among the 21–30 age group. This result is different from what was reported in Kuwait and the Middle East and North Africa (MENA) regional studies [15]. Similar profiles have been reported in other local studies, with the majority being the age group of 25 to 29 years [5,11].

In spite of the significant level of knowledge expressed in the study sample, the findings indicate a wide range of practices within the studied gym members. This suggested inadequate insight and a lack of healthcare professionals specializing in sports science who may be able to popularize an understanding of the wide range of adverse effects related to AAS use [7].

Our study revealed a negative association between monthly income and level of education with the prevalence of AAS use, which means that knowledge and education were independent of AAS use. Remarkably, these findings were also observed in previous studies that were performed in the region [5].

The most common types of AAS used in our findings were Anavar, Dianabol, and Deca Durabolin. This differed from previously reported data in another region of Saudi Arabi [17]. This could reflect the availability and preference for AAS in Saudi Arabia.

5. Conclusions

Based on this study, the prevalence of AAS users in Eastern Province, Saudi Arabia is high, which reflects the fact that gym members are at a higher risk of using AAS. Therefore, Regional Health Authorities in the Eastern Province region, Saudi Arabia, should urgently take measures to alleviate the potential adverse implications of AAS consumptions among young adults by using social media as an educational tool, or by distributing informative leaflets among AAS users.

6. Limitation

Although this is the first study conducted on gym members in the Eastern Province regarding AAS consumption, it had some limitations. Notably, it was only performed among male gym members, no blood work was included to investigate potential adverse implications of AAS, and finally, we used a self-reported questionnaire.

Author Contributions: W.A.; Y.E. and N.J. made substantial contributions to the conception, design, and acquisition of data. Y.E.; N.J. and A.A. helped in the analysis and interpretation of data. D.B.; and M.A.-H. drafted the manuscript. All authors have read and agreed to the published version of the manuscript.

Funding: The authors received no specific funding for this work.

Institutional Review Board Statement: The study was conducted according to the guidelines of the Declaration of Helsinki, and approved by the Institutional Review Board of Imam Abdulrahman Bin Faisal University (IRB-2018-01-174, 27 September 2018).

Informed Consent Statement: Informed consent was obtained from all subjects involved in the study.

Data Availability Statement: The data used to support the findings of this study are available from the corresponding authors upon reasonable request.

Conflicts of Interest: The authors declare that they have no competing interests.

References

1. Vorona, E.; Nieschlag, E. Adverse effects of doping with anabolic androgenic steroids in competitive athletics, recreational sports and bodybuilding. *Minerva Endocrinol.* **2018**, *43*, 476–488. [CrossRef] [PubMed]
2. Fisher, G.L.; Roget, N.A. *Encyclopedia of Substance Abuse Prevention, Treatment, and Recovery*; SAGE Publications: Thousand Oaks, CA, USA, 2009; pp. 72–75.
3. Hall, R.C.; Hall, R.C. Abuse of supraphysiologic doses of anabolic steroids. *South. Med. J.* **2005**, *98*, 550–555. [CrossRef]
4. Lang, T.; Streeper, T.; Cawthon, P.; Baldwin, K.; Taaffe, D.R.; Harris, T.B. Sarcopenia: Etiology, clinical consequences, intervention, and assessment. *Osteoporos Int.* **2010**, *21*, 543–549. [CrossRef] [PubMed]
5. Al-Harbi, F.; Gamaleddin, I.; Alsubaie, E.; Al-Surimi, K. Prevalence and Risk Factors Associated with Anabolic-androgenic Steroid Use: A Cross-sectional Study among Gym Users in Riyadh, Saudi Arabia. *Oman Med. J.* **2020**, *30*, e110. [CrossRef] [PubMed]
6. Piacentino, D.; Kotzalidis, G.D.; Del Casale, A.; Aromatario, M.R.; Pomara, C.; Girardi, P. Anabolic-androgenic Steroid use and Psychopathology in Athletes. *A Syst. Review. Curr Neuropharmacol.* **2015**, *13*, 101–121. [CrossRef] [PubMed]
7. Winkler, U.H. Effects of androgens on haemostasis. *Maturitas* **1996**, *24*, 147–155. [CrossRef]
8. Mhillaj, E.; Morgese, M.G.; Tucci, P.; Bove, M.; Schiavone, S.; Trabace, L. Effects of anabolic androgens on brain reward function. *Front. Neurosci.* **2015**, *9*, 295. [CrossRef] [PubMed]
9. Sando, B.G. Is it legal? Prescribing for the athlete. *Aust. Fam. Physician* **1999**, *28*, 549–553. [PubMed]
10. Logothetis, C.J. Organ preservation in bladder carcinoma: A matter of selection. *J. Clin. Oncol.* **1991**, *9*, 1525–1526. [CrossRef] [PubMed]
11. Bahri, A.; Mahfouz, M.S. Prevalence and awareness of anabolic androgenic steroid use among male body builders in Jazan, Saudi Arabia. *Trop. J. Pharm. Res.* **2017**, *16*, 1425–1430. [CrossRef]
12. Al Nozha, O.M.; Elshatarat, R.A. Influence of knowledge and beliefs on consumption of performance enhancing agents in north-western Saudi Arabia. *Ann. Saudi Med.* **2017**, *37*, 317–325. [CrossRef] [PubMed]
13. Tahtamouni, L.H.; Mustafa, N.H.; Alfaouri, A.A.; Hassan, I.M.; Abdalla, M.Y.; Yasin, S.R. Prevalence and risk factors for anabolic-androgenic steroid abuse among Jordanian collegiate students and athletes. *Eur. J. Public Health* **2008**, *18*, 661–665. [CrossRef]
14. Al-Falasi, O.; Al-Dahmani, K.; Al-Eisaei, K.; Al-Ameri, S.; Al-Maskari, F.; Nagelkerke, N.; Schneider, J. Knowledge, Attitude and Practice of Anabolic Steroids Use Among Gym Users in Al-Ain District, United Arab Emirates. *Open Sports Med.* **2008**, *2*, 75–81. [CrossRef]
15. Alsaeed, I.; Alabkal, J.R. Usage and perceptions of anabolic-androgenic steroids among male fitness centre attendees in Kuwait-a cross-sectional study. *Subst. Abus. Treat. Prev. Policy* **2015**, *10*, 1. [CrossRef]
16. Mullen, C.; Whalley, B.; Schifano, F.; Baker, J. Anabolic androgenic steroid abuse in the United Kingdom: An update. *Br. J. Pharmacol.* **2020**, *177*, 2180–2198. [CrossRef] [PubMed]
17. Jabari1, M.; Al-shehri, H. The prevalence of anabolic androgenic steroid use amongst athletes in Riyadh (Saudi Arabia). *Electron. Physician* **2016**, *25*, 3343–3347. [CrossRef] [PubMed]

Article

Effect of Heel-First Strike Gait on Knee and Ankle Mechanics

Shirin Aali [1], Farhad Rezazadeh [2], Georgian Badicu [3] and Wilhelm Robert Grosz [3,*]

1 Department of Sport Science Education, Farhangian University, Tehran 1998963341, Iran; shirin.aali1365@gmail.com
2 Department of Biomechanics and Sport Injuries, Kharazmi University of Tehran, Teheran 1998963341, Iran; rezazade.farhad@gmail.com
3 Department of Physical Education and Special Motricity, Faculty of Physical Education and Mountain Sports, Transilvania University of Brasov, 500068 Braşov, Romania; georgian.badicu@unitbv.ro
* Correspondence: wilhelm.grosz@unitbv.ro; Tel.: +40-722-741-729

Abstract: *Background and Objectives*: Acquiring knowledge about the magnitude and direction of induced joint forces during modifying gait strategies is critical for proper exercise prescription. The present study aimed to evaluate whether a heel-first strike pattern during gait can affect the biomechanical characteristics of ankle and knee joints among asymptomatic people. *Materials and Methods*: In this cross-sectional study performed in the biomechanics laboratory, 13 professional healthy male athletes walked on an instrumented walkway under two walking conditions. For the normal condition, subjects were instructed to walk as they normally would. For the heel-first strike condition, subjects were instructed to walk with heel-first strike pattern and increase heel contact duration as much as possible. Then, knee and ankle joint range of motions and moments, as well as vertical ground reaction force was measured by the Kistler force plate and Vicon motion analysis system. *Results*: Knee flexion angle at the initial contact and during stance phase was significantly lower when increasing the heel strike pattern. In addition, the mean values of the knee external rotation and adductor moments during heel strike condition were lower than those in normal walking. Further, the ankle dorsiflexion range of motion (ROM) during mid-stance increased significantly during heel-first strike pattern compared to the value in normal gait pattern. *Conclusions*: The modification of gait pattern including heel-first strike pattern can reduce the mechanical load applied to the knee, while improving the extensibility of gastro-soleus muscle complex.

Keywords: biomechanics; heel strike; ankle; gait change; gastrocnemius muscle stiffness

Citation: Aali, S.; Rezazadeh, F.; Badicu, G.; Grosz, W.R. Effect of Heel-First Strike Gait on Knee and Ankle Mechanics. *Medicina* **2021**, *57*, 657. https://doi.org/10.3390/medicina57070657

Academic Editor: Filipe Manuel Clemente

Received: 30 April 2021
Accepted: 23 June 2021
Published: 26 June 2021

Publisher's Note: MDPI stays neutral with regard to jurisdictional claims in published maps and institutional affiliations.

1. Introduction

Excessive joint forces play a significant role in developing musculoskeletal system impairment and pain which are the major causes of disability and form a major part of the high costs of health care in the industrialized world [1,2]. In particular, limited ankle dorsiflexion range of motion (ROM) is associated with many of the lower extremity injuries such as anterior cruciate ligament ACL, Achilles, and patella tendon injuries [3,4]. The relationship between gastro-soleus complex stiffness and many of the clinical problems of the foot and other joints of the kinetic chain are well-documented, which cause syndromes of the movement system [5,6].

Previous studies confirm the theories related to the existing mechanics of limited ankle dorsiflexion ROM and the injury of the kinetic chain joints [7,8]. The decreased ankle dorsiflexion may decrease anterior tibial translation at the ankle joint during stance phase of the gait cycle [7] and result in lowering the center of gravity during functional tasks such as walking [9]. This may compensate through mid-foot and subtalar joint pronation, knee joint flexion and valgus, which are related to chronic and acute injuries including ACL rupture, patella-femoral pain syndrome (PFPS), and knee osteoarthritis [10]. Some studies reported increased knee flexion and knee valgus during functional movements.

For example, walking among people with limited ankle dorsiflexion ROM supports this theory [8,11].

Further, limited ankle dorsiflexion ROM may increase the injury risk through changing the joint forces and muscle stiffness of the lower extremities. Accordingly, the reduced ankle dorsiflexion ROM and changes in knee and hip ROM resulted in increasing ground reaction forces, which are considered as the main causes of lower extremities injuries reported in the previous studies [12,13]. Another possibility is related to the relationship between limited ankle dorsiflexion and injuries through a series of mechanical compensatory patterns instead of a unique mechanical movement pattern in one joint. According to the dynamic system theory, they are considered as multiple biomechanical degrees of freedom recruiting variability as a common output. In addition, limited dorsiflexion may demonstrate loss of a biomechanical degree of freedom (movement pattern variability) which is related to various injuries in the overall kinetic chain [14,15].

Therefore, a good understanding of how gait strategies affect moment and joint forces may improve the results of therapeutic protocols. Acquiring knowledge about the magnitude and direction of induced joint forces during modifying gait strategies is critical for prescribing proper exercise. Modification may include a subtle change in lower-extremity position such as encouraging patients to walk with heel-first strike pattern and increasing heel contact duration as much as possible. Furthermore, it is documented that the musculoskeletal system is optimized subtly to minimize demanding stresses on bones and muscles; any improper change in the movement system, such as muscle imbalance or weakness, increases the joint forces significantly [16,17]. Thus, it is important to evaluate the effectiveness of joint forces from modifying gait patterns among people with gastrocnemius muscle stiffness.

Clinically speaking, the results of the previous studies indicated that the people with gastro-soleus muscle stiffness experienced increased peak foot pronation, knee valgus, magnitude of ground reaction forces, and knee adduction moment [18,19]. Therefore, it seems that changing the gait strategy with a simple guide, such as "walking with heel-first strike pattern and increasing heel contact duration as much as possible" during walking and the extensibility of gastro-soleus muscle complex is effective, especially for the patients with limited dorsiflexion and resulting pain in other joints of the kinetic chain (lower back pain) [17]. However, during the first step of designing a therapeutic protocol, it is unclear whether modifying gait pattern with a simple guide to "heel-first strike pattern" can reduce ground reaction forces, ankle dorsiflexion during the mid-stance phase of gait, knee flexion, knee external rotation, and adduction moment among the healthy people.

Thus, the present study aimed to investigate whether a heel-first strike pattern during gait can affect the selected kinetic and kinematic parameters of ankle and knee joint among healthy subjects. Principally, it is assumed that when a subject changes their gait pattern, the vertical ground reaction force, knee flexion angle during heel contact, mean ankle dorsiflexion during mid-stance, mean knee external rotation, and adduction moment are changed to reduce joint forces and the extensibility of the gastro-soleus muscle.

2. Materials and Methods

2.1. Participants

This cross-sectional study was performed on 13 professional competitive runners with five years of experience in the national athletics team (mean age = 25.2 $\pm$ 1.2, height = 179.9 $\pm$ 1.4 cm, weight = 76.6 $\pm$ 4.7 kg) which had normal dorsiflexion ROM in both legs. The mean ankle dorsiflexion AROM in the knee extension and flexion position was 14.4 $\pm$ 0.8 and 14.5 $\pm$ 0.9 degrees, respectively. The exclusion criteria included the history of trauma or ankle surgery, bone pathology, neurological disorders, and rheumatoid arthritis, inflammatory diseases, and any conditional abnormalities affecting the research process. The individuals recruited from the available community were selected by using the findings of a preliminary study to determine the sample size based on the variance of the test parameters among five participants.

Regarding the ethical consideration, all of the participants read and signed an approved informed consent letter before data collection. The study was confirmed by the Ethical Committee of the university. Participants had the right to decline to participate and withdraw from the research after initiating the process of data collection.

The dorsiflexion ROM was measured using a universal goniometer in both knee bending for soleus length, and extended knee for gastrocnemius length positions [16,17]. At least 10 degrees of ankle dorsiflexion are needed during the stance phase of the gait cycle, which can contribute to forward body movement for normal walking [5,6]. According to previous studies, the normal ankle dorsiflexion ROM should be ranged 10–15 degrees [20,21]. Therefore, at least 12 degrees was considered as the inclusion criterion for ankle dorsiflexion AROM.

2.2. Gait Testing Procedures

The participants underwent gait testing during a single data collection session. All procedures were carried out at Mowafaghian gait analysis laboratory of Sharif University of technology. Standard 9 mm retroreflective markers were placed over the anatomic landmarks including the heads of the first and fifth metatarsals, the posterior aspect of the calcaneus, and the medial and lateral malleoli, lateral knee joint lines, lateral epicondyle of femur, and anterior superior iliac spine, as well as laterally on shank and thigh of the affected foot determined by anatomic definitions from the Vicon Clinical Manager [19,22]. The position data of markers and ground reaction forces were processed with Vicon dynamic model using Plug-in-Gait-Workstation software version 4.3 (Oxford, UK) to generate the kinematic data and joint moment for sequential analysis.

In the next procedure, kinematic variables during walking were collected using a six-camera, motion capture system (Motion Analysis MX40S, VICON, Oxford, UK) sampling at 120 Hz. Then, the ground reaction force data were collected from two floor-mounted Kistler force plates (30 $\times$ 50, Winterthur, Switzerland) positioned in the middle of the 6 m walkway and sampling at 1200 Hz. Accordingly, three trials with clean force platform strikes were obtained for the dominant limb. All walking trials were performed barefoot based on the participants' self-selected and preferred walking speed.

Each participant accomplished three trials per each condition. The first is related to self-selected or preferred speed walking (normal walking). In the second condition, they were guided to increase the heel strike during walking (walking with heel-first strike pattern) based on the cue "walk with heel-first strike pattern and increase heel contact duration as much as possible". They were given only this simple instruction without any feedback during the study.

In addition, practice trials were allowed until the participants walked comfortably and could contact the force plate with only one foot without altering their gait. Typically, three trials are performed for each situation as a practice. The stance phase was determined from the moment the heel touched the force plate until the toe was lifted off.

2.3. Data Analysis

The data were collected from at least three trials for each condition with clean foot strikes from each foot. Then, the moments of force were calculated by mathematical equations including inverse dynamic method and link-segment model [3,6]. To calculate the knee adduction and rotation moment, the knee moment was first calculated on the frontal and horizontal planes, and then the time series data of moment-time were normalized. The first 60% of normalized time series was considered as the stance phase. According to ISB recommendations, the positive and negative moments are considered as knee external rotation and adduction torques, respectively [3,14]. Further, joint moments were normalized to body mass (NM/kg). Furthermore, temporal parameters such as step length, walking speed, and total stance time were identified for each trial. Finally, an independent t-test was used to compare group differences after determining the normal distribution of data using the Shapiro–Wilk test and the equality of variances between groups by using Levene's test. Statistical analysis was performed at the significance level of 0.05 and statistical power of 0.80 by SPSS software version 18.

3. Results

Table 1 indicates the mean and standard deviation for kinetic and kinematic variables, as well as spatiotemporal parameters of each group. Walking speed in normal walking is significantly greater than that of heel-first strike condition ($t = 2.21$, df = 12, $p = 0.05$). However, no significant difference was observed between the stance time and step length in two conditions ($t = -0.071$, df = 12, $p = 0.94$; $t = -1.25$, df = 12, $p = 0.24$, respectively). In addition, the mean values for the knee flexion-extension ROM during stance phase, knee flexion angle at initial contact, knee external rotation moment during stance phase were significantly higher in normal walking condition ($t = 2.26$, df = 12, $p = 0.02$; $t = 3.37$, df = 12, $p = 0.005$; $t = 3.09$, df = 12, $p = 0.03$, respectively). However, no significant difference was observed between the mean values for ankle dorsiflexion during stance phase in two conditions ($t = 0.22$, df = 12, $p = 0.82$). In addition, the mean ankle dorsiflexion during mid-stance was higher during walking with heel-first strike compared to normal walking ($t = -2.37$, df = 12, $p = 0.04$).

Table 1. The mean (standard deviation) for kinetic, kinematic, and spatiotemporal parameters.

Variable	Walking with Increased Heel Strike	Normal Walking	*p* Value	Cohen's D
Knee flexion-extension ROM during stance phase (degrees)	23.6 (7.3)	32.8 (12.7)	0.02 *	−1.34
Knee flexion angle at initial contact (degrees)	7.1 (3.7)	14.02 (6.4)	0.005 *	−1.32
Mean ankle dorsiflexion during stance phase (degrees)	7.9 (6.5)	5.6 (4.4)	0.82	0.41
Mean ankle dorsiflexion during mid-stance (degrees)	13.84 (2)	11.47 (1.5)	0.04 *	1.18
Mean knee external rotation moment during stance phase (NM/Kg)	0.18 (0.03)	0.23 (0.05)	0.03 *	−1.21
Mean knee adduction moment during stance phase (NM/kg)	4.01 (2.7)	7.6 (3.3)	0.01 *	−1.19
Total stance phase time (second)	0.79 (0.12)	0.79 (0.14)	0.94	0
Walking speed (m/s)	1.05 (0.02)	1.16 (0.16)	0.05	−0.96
Step length (m)	1.32 (0.11)	1.26 (0.08)	0.24	0.62

ROM: range of motion; * indicates statistical significance.

Table 2 indicates the mean and standard deviation of vertical ground-reaction force parameters. The mean value for the peak vertical force at the initial contact was significantly higher during walking with heel-first strike ($t = -2.88$, df = 12, $p = 0.01$). Further, no significant difference was reported between the mean value for peak vertical force during weight bearing in two conditions ($t = 0.78$, df = 12, $p = 0.45$). However, the mean value for peak vertical force during toe-off motion was greater during walking with heel-first strike than during the normal walking condition ($t = -4.38$, df = 12, $p = 0.002$).

Table 2. The mean (standard deviation) of ground reaction force parameters (data expressed as percentage of stance phase time).

Vertical Ground Reaction Force (N/kg)	Walking with Increased Heel Strike	Percentage of Stance Phase	Normal Walking	Percentage of Stance Phase	*p* Value	Cohen's D
The peak vertical force at the initial contact	0.53 (0.07)	0.8	0.35 (0.19)	1.2	0.01 *	−1.25
Peak vertical force during weight bearing	0.98 (0.06)	6	1.02 (0.16)	8	0.45	0.33
Peak vertical force during toe-off motion	1.09 (0.03)	20	1.05 (0.04)	24	0.002 *	−1.13

* indicates statistical significance.

4. Discussion

The results indicated that the ankle dorsiflexion ROM during the stance phase was greater in heel-first strike compared to the normal walking trial. Furthermore, the parameters of ground reaction force during walking with heel-first strike pattern were significantly higher than those in normal walking (Figure 1). The results could support the theory that heel-first strike during walking leads to a decrease in knee flexion angle at the initial contact and stance phase (Figure 2). In addition, two kinetic variables including the knee external rotation and adduction moment were significantly lower with increased heel-first strike trial (Figure 3). However, the spatiotemporal parameters such as step length and total stance phase time were not significantly different between trials, while walking speed in normal walking was significantly higher than walking with heel-first strike pattern condition.

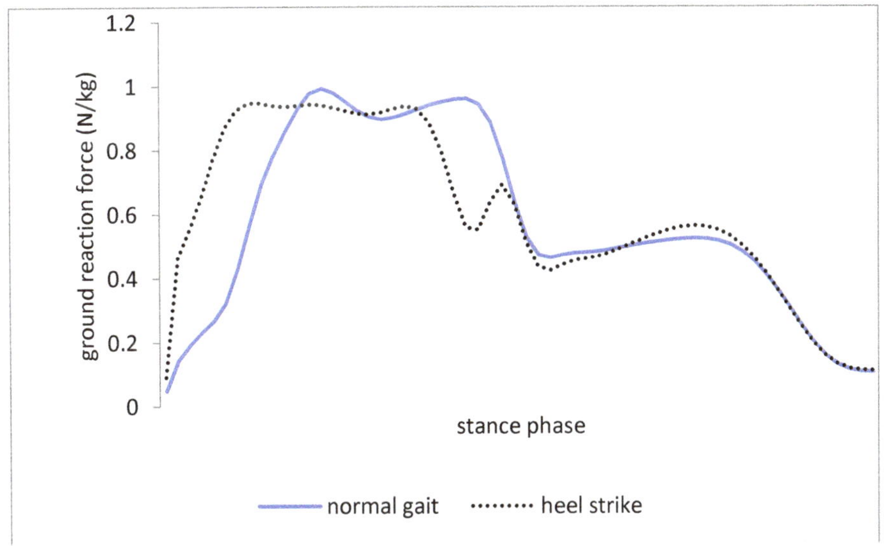

Figure 1. Ground reaction force during increased heel strike gait and normal gait during stance phase.

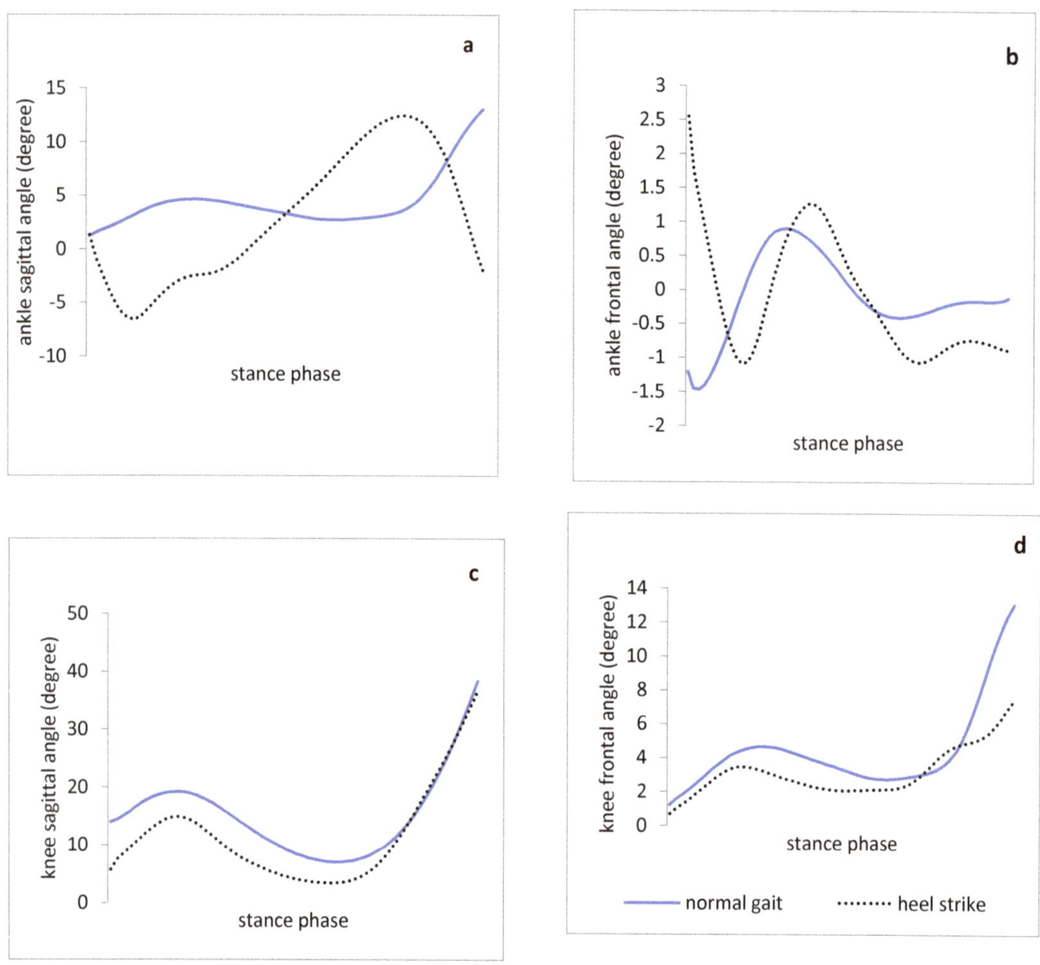

Figure 2. Joint angle patterns (degree) of healthy athletes with increased heel strike gait and normal gait during stance phase. (**a**,**b**): ankle joint angles in the sagittal and frontal plane, respectively. (**c**,**d**): knee joint angles in the sagittal and frontal planes, respectively.

Based on the results, the ground reaction forces, especially the peak impact with heel-first strike walking trial, were significantly greater than the normal walking trial (Figure 1). The findings are consistent with those of [23], which reported increased ground reaction forces in the athletes with the heel strike gait pattern. Grieve et al. [22,24] reported increased ground reaction forces due to the ankle kinematic changes, which are in line with the findings of the present study. In this study, changing the gait pattern with heel-first strike pattern resulted in changing the ground reaction force and knee joint kinematics. On the other hand, the walking speed in normal walking was significantly higher than walking with heel-first strike pattern condition (Table 1), which can be explained by increasing the heel contact duration in the walking with heel-first strike pattern condition rather than normal walking. In addition, walking with heel-first strike pattern changed the kinematics of the knee joint since a reduced inertia led to a compensatory change in the gait kinematics. However, the lack of studies in this area has created some difficulties for interpreting the results.

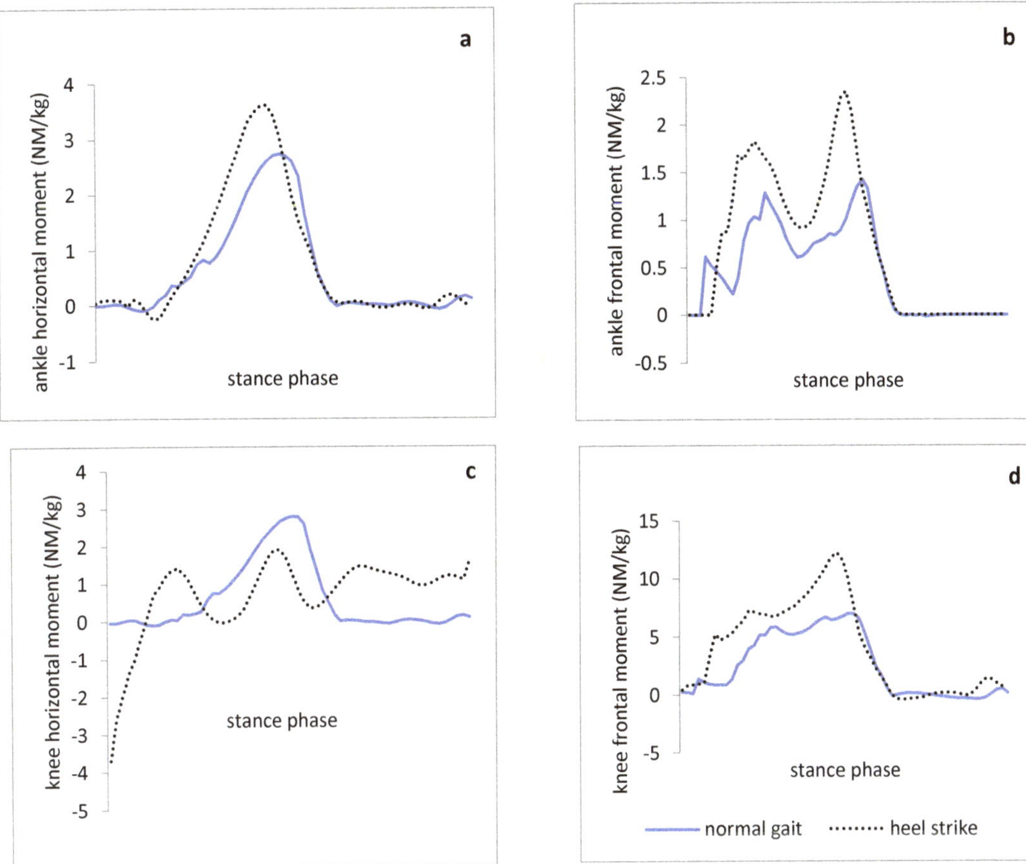

Figure 3. Joint moment pattern (N.M/Kg) of healthy athletes with increased heel strike gait and normal gait during stance phase. (**a**,**b**): ankle moment in the horizontal and frontal planes, respectively. (**c**,**d**): knee joint moment in the horizontal and frontal planes, respectively.

As displayed in Figure 2, knee flexion angle at the initial contact and the knee flexion ROM during the stance phase with heel-first strike pattern walking trial were significantly lower than those in the normal walking. Thus, the significance of the heel strike effect on ground reaction forces should be determined along with demonstrating the heel strike effect changing the knee joint kinematics in the sagittal plane. As a result, the knee and hip joints run into a more flexion position among the people with gastro-soleus stiffness compared to those in healthy people. This position puts the knee in the unlocked position leading to insufficient knee motor control, and compensatory movement patterns in the hip joint (recruiting increased hip extensor muscles and appearing to show synergistic dominance of the hamstring and gluteal muscles). Therefore, it seems that utilizing the cue "increase the heel contact duration" can change the knee flexion position during stance phase, and consequently, modify the insufficient motor control of the lower extremities.

It is believed that healthy people have a translation of tibia over the ankle joint during the stance phase of gait needed for forward propulsion. Despite the lack of significant difference of mean ankle dorsiflexion during stance phase between the trials, the mean ankle dorsiflexion ROM during mid-stance with heel-first strike pattern walking trial was significantly higher than that of normal walking (Figure 2). The finding of this pilot study may confirm that the heel-first strike walking pattern, as a functional pattern, can be effective in promoting the extensibility of the gastro-soleus muscle.

Furthermore, the values of the ground reaction force failed to decrease significantly in heel-first strike walking trial compared with those in normal walking. Thus, it is worth noting that gait strategies should be modified to achieve a special purpose, not as a set of positive effects. For example, increasing the ankle dorsiflexion ROM and decreasing the knee flexion angle for the individuals with gastro-soleus tightness are more essential than increasing ground reaction forces during walking with heel-first strike pattern since limited ankle dorsiflexion and increased knee flexion are considered as the risk factors for musculoskeletal injuries, such as patellofemoral pain syndrome, osteoarthritis, as well as synergistic dominance of hamstring muscles [10,15,16].

The results indicated that the mean external rotation and adduction moment of the knee in the heel-first strike pattern gait trial decreased significantly compared with those in a normal gait (Figure 3). In the present study, the knee external rotation and adduction moment were measured to identify and establish a movement pattern for the purpose of decreasing mechanical loading on the knee through the guidance to heel-first strike pattern since the improper loading of the knee is regarded as a risk factor for the knee osteoarthritis among the people with gastro-soleus tightness based on the literature [10,20]. In particular, medial knee compartment loading is considered as a clinical indication of the knee injuries and the knee external rotation and adduction moments are mentioned as an indirect measure of the medial knee compartment loading during functional tasks such as walking in most of the recent studies [20,21]. Therefore, early detection of the risk factors involved in knee osteoarthritis, as well as identifying the effective movement pattern for reducing the knee external rotation and adduction moment values may be more successful with the effectiveness of exercise interventions and preventing structural changes in the knee joint. Based on the results of the present study, walking with a heel-first strike pattern can reduce the loading forces of the knee joint. However, the findings cannot be generalized to those with gastro-soleus tightness. Therefore, further research is needed to ensure the effectiveness of heel-first strike gait pattern on reducing mechanical loading on the kinetic chain in people with gastro-soleus tightness.

5. Conclusions

Based on the results, heel-first strike gait pattern in healthy athletes could make kinematic changes in the knee and ankle joints in all three movement planes. For example, knee flexion angle at the initial contact and the knee flexion ROM during the stance phase decreased with heel-first strike pattern walking. Additionally, some kinetic changes including the mean values of the knee external rotation and adductor moments during heel strike condition were lower than those in normal walking. In addition, heel-first strike gait pattern with increased ankle dorsiflexion at the heel contact led to the extensibility of gastro-soleus muscle complex.

The main limitation is that the participants had no limited dorsiflexion ROM. Thus, the people with gastro-soleus tightness failed to respond to heel-first strike gait pattern like normal people. The present pilot study aimed to evaluate the concept of heel-first strike gait pattern for decreasing knee flexion, increasing ankle dorsiflexion, minimizing knee external rotation and adduction moment during gait, and applying the pattern in people with gastro-soleus tightness. By confirming the heel-first strike gait pattern effect on kinematic changes in this study, another study can be conducted on applying heel strike gait in the athletes with gastro-soleus tightness.

Author Contributions: S.A. and F.R. conceived and designed the experiments; S.A. performed the experiments; F.R., G.B. and W.R.G. analyzed the data; S.A. and F.R. contributed reagents and materials; S.A., F.R., G.B. and W.R.G. wrote the paper. All authors have read and agreed to the published version of the manuscript.

Funding: This research did not receive any specific grant from funding agencies in the public, commercial, or not-for-profit sectors.

Institutional Review Board Statement: The study was conducted according to the guidelines of the Declaration of Helsinki, and approved by the Ardabil University of Medical Sciences (protocol code: IR.ARUMS.REC.1400.95 and date of approval: 10 November 2020).

Informed Consent Statement: Informed consent was obtained from all subjects involved in the study.

Data Availability Statement: The data presented in this study are available on request from the author Farhad Rezazadeh.

Acknowledgments: Thanks to numerous individuals who participated in this study.

Conflicts of Interest: The authors declare no conflict of interest.

References

1. Mavčič, B.; Slivnik, T.; Antolič, V.; Iglič, A.; Kralj-Iglič, V. High contact hip stress is related to the development of hip pathology with increasing age. *Clin. Biomech.* **2004**, *19*, 939–943. [CrossRef]
2. Dixon, J.B. Gastrocnemius vs. soleus strain: How to differentiate and deal with calf muscle injuries. *Curr. Rev. Musculoskelet. Med.* **2009**, *2*, 74–77. [CrossRef]
3. Mason-Mackay, A.; Whatman, C.; Reid, D. The effect of reduced ankle dorsiflexion on lower extremity mechanics during landing: A systematic review. *J. Sci. Med. Sport* **2017**, *20*, 451–458. [CrossRef]
4. Didier, J.J.; West, V.A. Vertical Jumping and Landing Mechanics: Female Athletes and Nonathletes. *Int. J. Athl. Ther. Train.* **2011**, *16*, 17–20. [CrossRef]
5. You, J.-Y.; Lee, H.-M.; Luo, H.-J.; Leu, C.-C.; Cheng, P.-G.; Wu, S.-K. Gastrocnemius tightness on joint angle and work of lower extremity during gait. *Clin. Biomech.* **2009**, *24*, 744–750. [CrossRef]
6. Yoon, J.-Y.; Hwang, Y.-I.; An, D.-H.; Oh, J.-S. Changes in Kinetic, Kinematic, and Temporal Parameters of Walking in People With Limited Ankle Dorsiflexion: Pre-Post Application of Modified Mobilization With Movement Using Talus Glide Taping. *J. Manip. Physiol. Ther.* **2014**, *37*, 320–325. [CrossRef] [PubMed]
7. Piva, S.R.; Goodnite, E.A.; Childs, J.D. Strength Around the Hip and Flexibility of Soft Tissues in Individuals With and Without Patellofemoral Pain Syndrome. *J. Orthop. Sports Phys. Ther.* **2005**, *35*, 793–801. [CrossRef]
8. Mauntel, T.C.; Begalle, R.L.; Cram, T.R.; Frank, B.S.; Hirth, C.J.; Blackburn, T.; Padua, D. The Effects of Lower Extremity Muscle Activation and Passive Range of Motion on Single Leg Squat Performance. *J. Strength Cond. Res.* **2013**, *27*, 1813–1823. [CrossRef]
9. Macrum, E.; Bell, D.R.; Boling, M.; Lewek, M.; Padua, D. Effect of Limiting Ankle-Dorsiflexion Range of Motion on Lower Extremity Kinematics and Muscle-Activation Patterns During a Squat. *J. Sport Rehabil.* **2012**, *21*, 144–150. [CrossRef] [PubMed]
10. Paoloni, M.; Mangone, M.; Fratocchi, G.; Murgia, M.; Saraceni, V.M.; Santilli, V. Kinematic and kinetic features of normal level walking in patellofemoral pain syndrome: More than a sagittal plane alteration. *J. Biomech.* **2010**, *43*, 1794–1798. [CrossRef] [PubMed]
11. Bell, D.R.; Vesci, B.J.; DiStefano, L.J.; Guskiewicz, K.M.; Hirth, C.J.; Padua, D.A. Muscle Activity and Flexibility in Individuals With Medial Knee Displacement During the Overhead Squat. *Athl. Train. Sports Health Care* **2012**, *4*, 117–125. [CrossRef]
12. Wang, S.S.; Whitney, S.L.; Burdett, R.G.; Janosky, J.E. Lower Extremity Muscular Flexibility in Long Distance Runners. *J. Orthop. Sports Phys. Ther.* **1993**, *17*, 102–107. [CrossRef] [PubMed]
13. Norcross, M.F.; Lewek, M.; Padua, D.; Shultz, S.J.; Weinhold, P.; Blackburn, J.T. Lower Extremity Energy Absorption and Biomechanics During Landing, Part I: Sagittal-Plane Energy Absorption Analyses. *J. Athl. Train.* **2013**, *48*, 748–756. [CrossRef]
14. Guccione, A.A.; Neville, B.T.; George, S. Optimization of Movement: A Dynamical Systems Approach to Movement Systems as Emergent Phenomena. *Phys. Ther.* **2018**, *99*, 3–9. [CrossRef] [PubMed]
15. Hamill, J.; Palmer, C.; Van Emmerik, R.E.A. Coordinative variability and overuse injury. *Sports Med. Arthrosc. Rehabil. Ther. Technol.* **2012**, *4*, 45. [CrossRef] [PubMed]
16. Lewis, C.L.; Garibay, E.J. Effect of increased pushoff during gait on hip joint forces. *J. Biomech.* **2015**, *48*, 181–185. [CrossRef]
17. Sahrmann, S. *Movement System Impairment Syndromes of the Extremities, Cervical and Thoracic Spines*; Elsevier: Amsterdam, The Netherlands; Mosby: Maryland Heights, MO, USA, 2011.
18. Kuhman, D.; Paquette, M.R.; Peel, S.A.; Melcher, D.A. Comparison of ankle kinematics and ground reaction forces between prospectively injured and uninjured collegiate cross country runners. *Hum. Mov. Sci.* **2016**, *47*, 9–15. [CrossRef]
19. Noehren, B.; Schmitz, A.; Hempel, R.; Westlake, C.; Black, W. Assessment of strength, flexibility, and running mechanics in men with iliotibial band syndrome. *J. Orthop. Sports Phys. Ther.* **2014**, *44*, 217–222. [CrossRef]
20. Meireles, S.; De Groote, F.; Reeves, N.; Verschueren, S.; Maganaris, C.; Luyten, F.; Jonkers, I. Knee contact forces are not altered in early knee osteoarthritis. *Gait Posture* **2016**, *45*, 115–120. [CrossRef] [PubMed]
21. Mündermann, A.; Asay, J.; Mündermann, L.; Andriacchi, T.P. Implications of increased medio-lateral trunk sway for ambulatory mechanics. *J. Biomech.* **2008**, *41*, 165–170. [CrossRef]

22. Grieve, R.; Barnett, S.; Coghill, N.; Cramp, F. Myofascial trigger point therapy for triceps surae dysfunction: A case series. *Man. Ther.* **2013**, *18*, 519–525. [CrossRef] [PubMed]
23. Mercer, J.A.; Horsch, S. Heel–toe running: A new look at the influence of foot strike pattern on impact force. *J. Exerc. Sci. Fit.* **2015**, *13*, 29–34. [CrossRef] [PubMed]
24. Grieve, R.; Cranston, A.; Henderson, A.; John, R.; Malone, G.; Mayall, C. The immediate effect of triceps surae myofascial trigger point therapy on restricted active ankle joint dorsiflexion in recreational runners: A crossover randomised controlled trial. *J. Bodyw. Mov. Ther.* **2013**, *17*, 453–461. [CrossRef] [PubMed]

 *medicina* 

Article

In-Season Internal and External Workload Variations between Starters and Non-Starters—A Case Study of a Top Elite European Soccer Team

Rafael Oliveira [1,2,3,*], Luiz H. Palucci Vieira [4], Alexandre Martins [1,2], João Paulo Brito [1,2,3], Matilde Nalha [1], Bruno Mendes [5] and Filipe Manuel Clemente [6,7]

1 Sports Science School of Rio Maior–Polytechnic, Institute of Santarém, 2140-413 Rio Maior, Portugal; alexandremartins@esdrm.ipsantarem.pt (A.M.); jbrito@esdrm.ipsantarem.pt (J.P.B.); matildenalha@gmail.com (M.N.)
2 Life Quality Research Centre, 2140-413 Rio Maior, Portugal
3 Research Center in Sport Sciences, Health Sciences and Human Development, Quinta de Prados, Edifício Ciências de Desporto, 5001-801 Vila Real, Portugal
4 Graduate Program in Movement Sciences, MOVI-LAB Human Movement Research Laboratory, Physical Education Department, School of Sciences, UNESP São Paulo State University, 17033-360 Bauru, Brazil; luiz.palucci@unesp.br
5 Falculty of Human Kinetics, University of Lisboa, 1649-002 Lisboa, Portugal; brunomendes94@hotmail.com
6 Escola Superior Desporto e Lazer, Instituto Politécnico de Viana do Castelo, Rua Escola Industrial e Comercial de Nun'Álvares, 4900-347 Viana do Castelo, Portugal; filipe.clemente5@gmail.com
7 Instituto de Telecomunicações, Delegação da Covilhã, 1049-001 Lisboa, Portugal
* Correspondence: rafaeloliveira@esdrm.ipsantarem.pt

Citation: Oliveira, R.; Palucci Vieira, L.H.; Martins, A.; Brito, J.P.; Nalha, M.; Mendes, B.; Clemente, F.M. In-Season Internal and External Workload Variations between Starters and Non-Starters—A Case Study of a Top Elite European Soccer Team. *Medicina* **2021**, 57, 645. https://doi.org/10.3390/medicina57070645

Academic Editor: José Antonio de Paz Fernández

Received: 3 May 2021
Accepted: 21 June 2021
Published: 23 June 2021

Abstract: *Background and Objectives:* Interpretation of the load variations across a period seems important to control the weekly progression or variation of the load, or to identify in-micro- and mesocycle variations. Thus, the aims of this study were twofold: (a) to describe the in-season variations of training monotony, training strain and acute:chronic workload ratio (ACWR) through session ratings of perceived exertion (s-RPE), total distance and high-speed running (HSR); and (b) to compare those variations between starters and non-starters. *Materials and Methods:* Seventeen professional players from a European First League team participated in this study. They were divided in two groups: starters ($n = 9$) and non-starters ($n = 8$). The players were monitored daily over a 41-week period of competition where 52 matches occurred during the 2015–2016 in-season. Through the collection of s-RPE, total distance and HSR, training monotony, training strain and ACWR were calculated for each measure, respectively. Data were analyzed across ten mesocycles (M: 1 to 10). Repeated measures ANOVA was used with the Bonferroni post hoc test to compare M and player status. *Results:* The results revealed no differences between starters vs. non-starters ($p > 0.05$). M6 had a greater number of matches and displayed higher values for monotony (s-RPE, total distance and HSR), strain (only for total distance) and ACWR (s-RPE, TD and HSR). However, the variation patterns for all indexes displayed some differences. *Conclusions:* The values of both starters and non-starters showed small differences, thus suggesting that the adjustments of training workloads that had been applied over the season helped to reduce differences according to the player status. Even so, there were some variations over the season (microcycles and mesocycles) for the whole team. This study could be used as a reference for future coaches, staff and scientists.

Keywords: acute/chronic workload ratio; high-speed running; in-season; non-starters; RPE; soccer; starters; training monotony; training strain

1. Introduction

Monitoring of the training load in soccer has become popular, whereby two main dimensions of load are considered [1]: (i) internal and (ii) external. The external load can be considered as the physical demands that occur in the players in response to the implemented drill/task, while the internal load corresponds to the psychophysiological responses to the external load [2]. Different outcomes can be considered for each of the dimensions, although the rate of perceived exertion (RPE) and heart rate responses are the most used measures associated with internal load [3]. On the other hand, in soccer, the external load is typically characterized by the distance covered at different speed thresholds, or the inertial-derived measures such as accelerations/decelerations or composite variables (e.g., player load) [4].

Monitoring loads allows one to identify the consequence of training plans on the players and to individualize the analysis [5]. Although it is useful to look for accurate measures representing the impact in a training session [6], interpretation of the load variations across a period of time also seems to be important [7]. In fact, calculating workload measures is a part of the strategies to control the weekly progression or variation of the load, or to identify within-week variations [8]. Among the possibilities, acute load (representing the accumulated load during a week), chronic load (typically represented by the mean load in the past weeks), acute:chronic workload ratio (ACWR, representing the relationship between acute and chronic workloads) [9], training monotony (TM) (representing the variability of load within the week) and training strain (TS) (representing the variability of the load multiplied by the acute load) [10] are some examples of how to control load taking into consideration different measures.

Considering that some of these measures are sensitive to load fluctuations, it can be expected that participating or not participating in soccer matches may influence the workload measures reported for the players. For example, it is expectable that players with greater participation in matches present greater values of accumulated load and chronic load. However, as a consequence, players with less participation should be carefully managed to be prepared for participating in matches and coping with a spike in load. Despite the apparently obvious consequence of participating more or less in matches being related with different workload measures, reports on this matter are limited [11]. For example, similar comparisons between starters and non-starters regarding the workload measures of new body load and metabolic power were found [11]. In junior soccer players, it was also found that weekly internal and external load measures were also significantly greater in starters than in substitute players [12].

However, the above-mentioned results still need more research that provides some description about the workload measures' variations in accordance with the level of participation of players in elite soccer. This should be further researched to provide information about how to manage players with match stimulus and to identify possible strategies to level the load with individualized training for those who are not playing. Based on that, the aims of this study were twofold: (a) to describe the in-season variations of TM, TS and ACWR through s-RPE, total distance and high-speed running (HSR); and (b) to compare those variations between starters and non-starters.

2. Materials and Methods

2.1. Subjects

Seventeen elite soccer players participated in this study. The players belong to a team that participated in the UEFA Champions League. They were divided into two groups: starters (n = 9, age 26.2 ± 3.5 years, 180.1 ± 6.8 cm and 78.7 ± 5.8 kg) and non-starters (n = 8, 24.5 ± 4.6 years, 182. ± 6.8 cm and 76.6 ± 4.3 kg). The inclusion criteria were regular participation in most of the training sessions (80% of weekly training sessions), while the exclusion criteria included lack of player information, illness and/or injury for two consecutive weeks. Goalkeepers were excluded from the study. The criteria to define starters and non-starters were assessed week by week against a player´s attendance time at the match and training sessions, and to be considered a starter, a player had to complete at least 60 minutes in three consecutive matches; players who did not achieve this duration were considered non-starters [13]. All participants were familiarized with the training protocols and signed informed consent prior to the investigation. This study was conducted according to the requirements of the Declaration of Helsinki and was approved by the Ethics Committee of Polytechnic Institute of Santarém (252020Desporto).

2.2. Design

Training load data were collected over a 41-week competition period, in which 52 matches occurred during the 2015–2016 in-season. The team used for data collection competed in four official competitions across the season, including the UEFA Champions League, the national league and two more national cups from their own country. For the purposes of the present study, all of the sessions carried out as the main team sessions were considered. This refers to training sessions in which both the starting and non-starting players trained together. Only data from training sessions were considered. Data from rehabilitation or additional training sessions of recuperation were excluded. This means that sessions after the match day were included whenever both starters and non-starters trained together, but other kinds of recovery training were excluded. This study did not influence or alter the training sessions in any way. Training data collection for this study was carried out at the soccer club's outdoor training pitches. Total minutes of training sessions included the warm-up, main phase and slow-down phase plus stretching.

The season was organized into 10 mesocycles (M: 1–10). The number of training sessions, number of competitive matches and total training duration for starters and non-starters are presented in Table 1.

Table 1. Training sessions and number of competitive matches during the 41-week period.

Mesocycle (M)	**M1**	**M2**	**M3**	**M4**	**M5**	**M6**	**M7**	**M8**	**M9**	**M10**
Training sessions (*n*)	16	20	18	18	20	20	19	20	18	20
Session duration, total minutes, ST	1501	1778	986	1495	1062	864	1410	1519	1206	1227
Session duration, total minutes, NST	1585	1832	1029	1424	1197	1272	1599	1441	1358	1382
Number of matches (*n*)	4	5	4	5	6	8	5	4	7	4

ST = Starters; NST = Non-starters.

2.3. Internal Training Load Quantification

During training sessions, the CR10-point scale, adapted by Foster et al. was applied [14]. Specifically, thirty minutes after the end of each training session, players rated their RPE value using an app on a tablet. The scores provided by the players were then multiplied by the training duration to obtain the s-RPE [14,15]. The players were previously familiarized with the scale, and all answers were provided individually to avoid non-valid scores.

2.4. External Training Load Quantification

Global positioning system (GPS) units (Viper pod 2, STATSports, Belfast, UK) with 10 Hz frequency were used to monitor the training duration, total distance and HSR (above 19 km/h) for each player. For better satellite reception of the GPS antenna, the GPS unit was placed on the upper back between the left and right scapula through a custom-made vest. Previously, Beato et al. [16] positively tested the validity and reliability of linear, multidirectional and soccer-specific activities through this system. Thirty minutes before the start of a training session, all devices were turned on to acquire satellite signals and to provide synchronization between the GPS clock and the satellite's atomic clock. After the training sessions, the Viper PSA software (STATSports, Belfast, UK) was used to download data and to clip the entire training session (i.e., from the beginning of the warm-up to the end of the last organized drill). In order to avoid inter-unit error, players wore the same GPS device in each training session.

2.5. Calculations of Training Indexes

Through s-RPE, total distance and HSR, the following variables were calculated: (i) TM (mean of training load during the seven days of the week divided by the standard deviation of the training load of the seven days) [11,17], (ii) TS (sum of the training loads for all training sessions during a week multiplied by training monotony) [11,17] and (iii) ACWR (dividing the acute workload, i.e., the 1-week rolling workload data, by the chronic workload, i.e., the rolling 4-week average workload data) [18–22].

2.6. Statistical Analysis

Data were analyzed using SPSS version 22.0 (SPSS Inc., Chicago, IL, USA) for Windows. Initially, descriptive statistics were used to describe and characterize the sample. The Shapiro–Wilk and Levene tests were used to test the assumption of normality and homoscedasticity, respectively. A repeated measures ANOVA was used with the Bonferroni post hoc test once variables obtained normal distribution (Shapiro–Wilk > 0.05), and the Friedman and Mann–Whitney tests were used for variables that did not obtain normal distribution in order to compare different M and groups. Hedge's g effect size (95% confidence interval) was also calculated. Hopkins' thresholds for effect size statistics were used, as follows: $\leq$0.2, trivial; >0.2, small; >0.6, moderate; >1.2, large; >2.0, very large; and >4.0, nearly perfect [21]. Results were considered significant with $p \leq 0.05$.

3. Results

Figures 1–3 show an overall view of the weekly average for TM, TS and ACWR calculated through the s-RPE, total distance and HSR across the in-season for starter and non-starter players. Overall, Figure 1 shows that the highest $TM_{s\text{-}RPE}$ occurred in week 1 for both starters and non-starters (7.2 and 7.0 AU, respectively), while the lowest value occurred in week 19 for starters (1.5 AU) and week 2 for non-starters (1.5 AU). The highest $TS_{s\text{-}RPE}$ occurred in week 41 for both starters (8498.0 AU) and non-starters (15,263.9 AU), while the lowest values occurred in week 30 for starters (110.2 AU) and week 19 for non-starters (1310.9 AU). The highest $ACWR_{s\text{-}RPE}$ occurred in week 21 for starters (1.6 AU) and week 10 for non-starters (1.5 AU), while the lowest $ACWR_{s\text{-}RPE}$ occurred in week 36 for starters (0.5 AU) and week 17 for non-starters (0.7).

Figure 2 shows that the highest TM_{TD} occurred in week 21 for both starters and non-starters (38.2 and 17.1 AU, respectively), while the lowest values occurred in week 2 for both starters and non-starters (2.0 and 1.9 AU, respectively). The highest TS_{TD} occurred in week 21 for starters (558,935.0 AU) and week 15 for non-starters (282,938.6 AU), while the lowest values occurred in week 36 for starters (35,441 AU) and non-starters (42,676.8 AU). The highest $ACWR_{TD}$ occurred in week 10 for both starters (1.6 AU) and non-starters (1.6 AU), and the lowest $ACWR_{TD}$ occurred in week 36 for both starters (0.7 AU) and non-starters (0.8 AU).

Figure 3 shows that the highest TM_{HSR} occurred in week 21 for starters (2.9 AU) and week 36 for non-starters (2.9 AU), while the lowest values occurred in week 20 for starters (0.7 AU) and week 39 for non-starters (0.8 AU). The highest TS_{HSR} occurred in week 4 for starters (3855.6 AU) and week 10 for non-starters (3578.0 AU), while the lowest values occurred in week 18 for starters (218.1 AU) and week 14 for non-starters (365.8 AU). The highest $ACWR_{HSR}$ occurred in week 10 for both starters (1.6 AU) and non-starters (1.6 AU), while the lowest $ACWR_{HSR}$ values occurred in week 9 for starters (0.4 AU) and week 4 for non-starters (0.4 AU).

Table 2 presents the average values and differences between starters and non-starters during the 10 mesocycles for all variables analyzed. There are no differences between the groups.

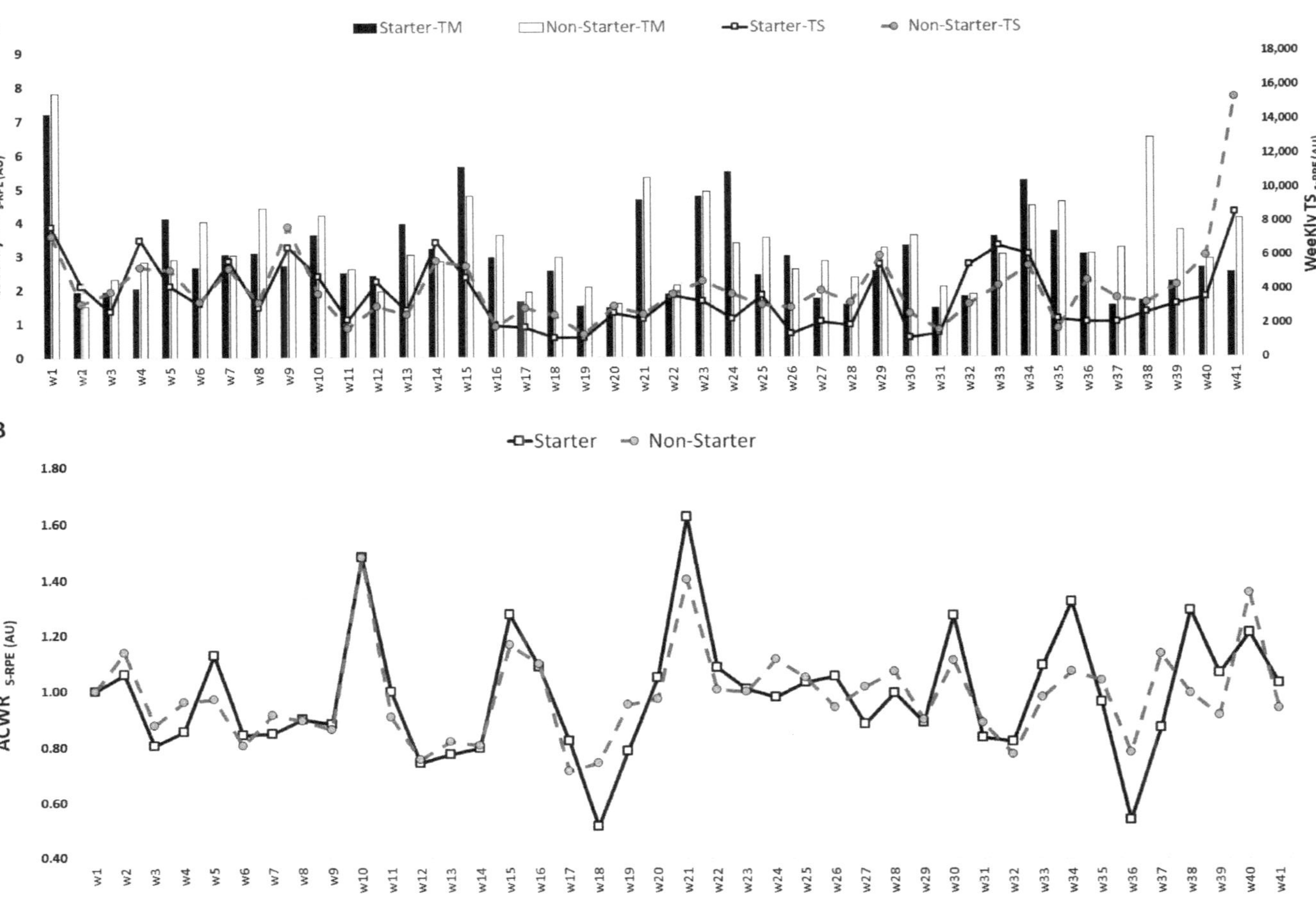

Figure 1. TM, TS (**A**) and ACWR (**B**) variations calculated through the s-RPE across 41 weeks for starters and non-starters.

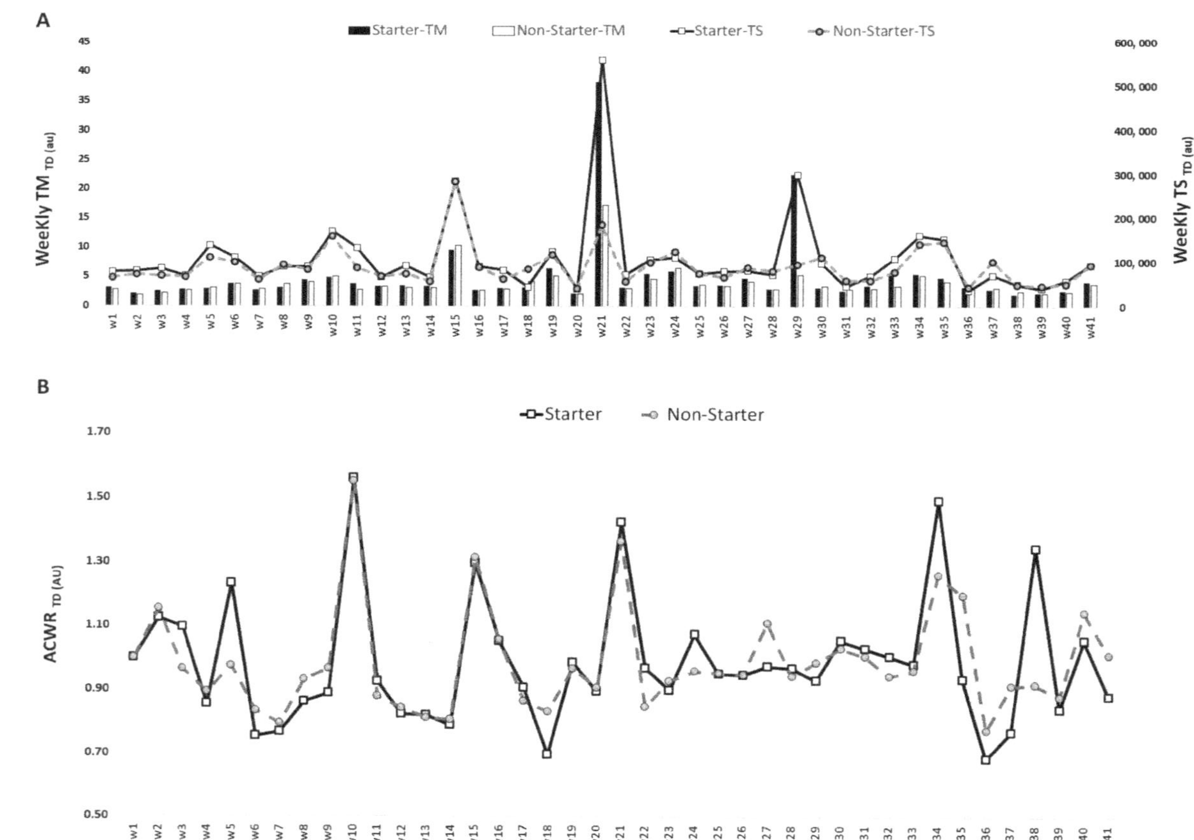

Figure 2. TM, TS (**A**) and ACWR (**B**) variations calculated through the total distance across 41 weeks for starters and non-starters.

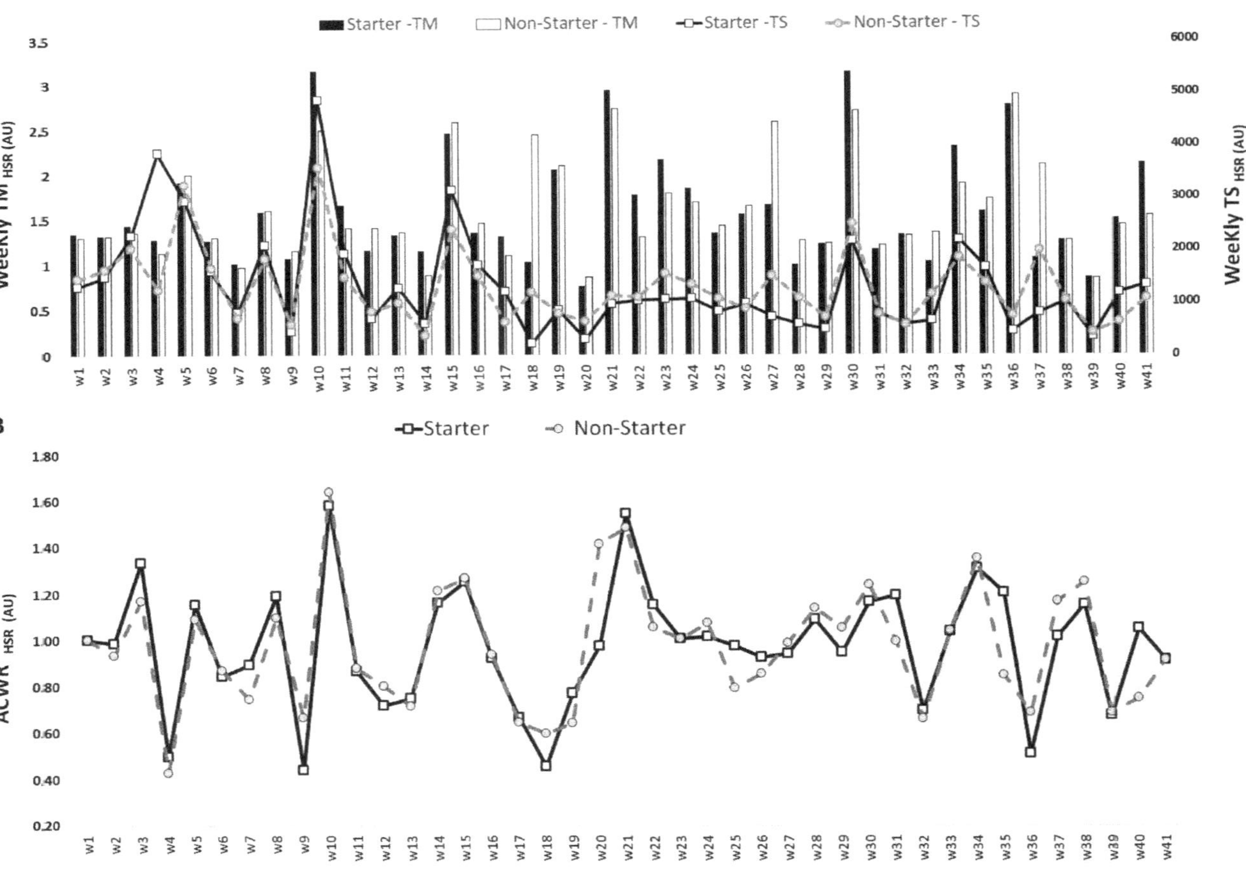

Figure 3. TM, TS (**A**) and ACWR (**B**) variations calculated through the HSR across 41 weeks for starters and non-starters.

Table 2. Differences between starters and non-starters during the 10 mesocycles, mean ± SD.

Variables	M1	M2	M3	M4	M5	M6	M7	M8	M9	M10
TM s-RPE (AU), ST	3.3 ± 1.1	3.2 ± 0.8	2.8 ± 0.2	3.9 ± 0.4	1.8 ± 0.3	4.4 ± 1.0	2.2 ± 0.3	2.3 ± 0.3	2.5 ± 0.2	3.3 ± 0.7
TM s-RPE (AU), NST	3.6 ± 1.2	3.6 ± 0.8	3 ± 0.3	3.5 ± 0.4	2.1 ± 0.3	3.8 ± 1.1	2.8 ± 0.3	2.7 ± 0.4	2.6 ± 0.3	3.8 ± 0.7
TS s-RPE (AU), ST	5370.6 ± 881.1	3972.5 ± 900.1	4454.0 ± 510.3	4002.1 ± 445.4	1522.6 ± 486.3	2839.1 ± 505.2	2220.8 ± 367.0	2442.1 ± 538.0	3334.9 ± 667.1	4202.4 ± 949.1
TS s-RPE (AU), NST	5101.1 ± 934.5	4206.7 ± 954.7	4035.9 ± 541.2	3764.3 ± 472.4	2274.6 ± 512.6	3593.6 ± 535.9	3268.3 ± 389.3	8308.2 ± 570.6	3290.7 ± 707.5	3933.4 ± 1006.7
ACWR s-RPE(AU), ST	0.9 ± 0.02	0.9 ± 0.03	1.0 ± 0.01	1.0 ± 0.03	0.8 ± 0.03	1.2 ± 0.04	1.0 ± 0.03	1.0 ± 0.03	1.0 ± 0.04	0.9 ± 0.1
ACWR s-RPE (AU), NST	1.0 ± 0.03	0.9 ± 0.03	1.0 ± 0.01	1.0 ± 0.04	0.8 ± 0.03	1.1 ± 0.05	1.0 ± 0.04	0.9 ± 0.03	0.9 ± 0.04	1.0 ± 0.05
TM TD (AU), ST	2.7 ± 0.1	3.2 ± 0.1	4.1 ± 0.2	4.8 ± 0.3	3.7 ± 0.4	12.3 ± 2.7	3.7 ± 0.1	8.0 ± 2.5	3.6 ± 0.2	4.1 ± 0.3
TM TD (AU), NST	2.4 ± 0.2	3.3 ± 0.1	3.8 ± 0.3	4.8 ± 0.3	3.6 ± 0.4	7.2 ± 2.9	3.5 ± 0.1	3.7 ± 2.6	3.2 ± 0.3	3.8 ± 0.4
TS TD (AU), ST	76,836.5 ± 3760.5	100,533.8 ± 5541.1	113,493.5 ± 6692.5	132,192.8 ± 10097.1	71,403.2 ± 7200.7	199,545.0 ± 39571.2	75,732.0 ± 3461.4	127,443.4 ± 30,416.6	79,449.1 ± 5330.6	104,429.4 ± 9679.7
TS TD (AU), NST	66,845.5 ± 3988.6	92,677.9 ± 5877.2	97,736.2 ± 798.5	124,250.7 ± 10709.5	75,171.9 ± 7637.5	127,445.4 ± 41971.6	76,288.8 ± 3671.3	80,471.7 ± 32,261.7	76,347.3± 5653.9	107,630.7 ± 10,266.9
ACWR TD (AU), ST	1.0 ± 0.02	0.9 ± 0.03	1.1 ± 0.03	1.0 ± 0.02	0.9 ± 0.01	1.1 ± 0.03	1.0 ± 0.02	1.0 ± 0.01	1.0 ± 0.01	1.0 ± 0.03
ACWR TD (AU), NST	1.0 ± 0.02	0.9 ± 0.03	1.0 ± 0.03	1.0 ± 0.02	1.0 ± 0.01	1.1 ± 0.03	1.0 ± 0.02	1.0 ± 0.01	1.0 ± 0.01	1.0 0.03
TM HSR (AU), ST	1.3 ± 0.06	1.4 ± 0.1	1.8 ± 0.1	1.6 ± 0.1	1.3 ± 0.3	2.1 ± 0.1	1.4 ± 0.1	1.7 ± 0.1	1.7 ± 0.2	1.9 ± 0.3
TM HSR (AU), NST	1.3 ± 0.1	1.4 ± 0.1	1.6 ± 0.1	1.6 ± 0.1	1.6 ± 0.3	1.8 ± 0.1	1.7 ± 0.1	1.6 ± 0.2	1.6 ± 0.2	2.2 ± 1.4
TS HSR (AU), ST	2226.2 ± 482.7	1857.1 ± 288.3	2051.2 ± 308.4	1676.5 ± 280.9	641.0 ± 205.1	1008.5 ± 185.2	768.6 ± 145.6	1003.7 ± 179.8	1044.7 ± 212.0	1269.6 ± 165.9
TS HSR (AU), NST	1586.2 ± 512.0	1806.5 ± 305.8	1617.6 ± 327.2	1310.9 ± 297.9	811.9 ± 217.5	1253.6 ± 196.5	1132.8 ± 154.4	1140.1 ± 190.7	1245.5 ± 224.9	1479.0 ± 176.0
ACWR HSR (AU), ST	1.0 ± 0.04	1.0 ± 0.04	0.9 ± 0.03	1.0 ± 0.05	0.7 ± 0.06	1.2 ± 0.08	1.0 ± 0.05	1.0 ± 0.3	1.0 ± 0.3	1.1 ± 0.04
ACWR HSR (AU), NST	0.9 ± 0.04	1.0 ± 0.05	1.0 ± 0.04	1.0 ± 0.06	0.8 ± 0.06	1.2 ± 0.09	0.9 ± 0.05	1.0 ± 0.03	1.0 ± 0.03	1.0 ± 0.05

M = mesocycle; RPE = rating of perceived exertion; s-RPE = session rating of perceived exertion; TM = training monotony; TS = training strain; ACWR = acute:chronic workload ratio; AU = arbitrary units; ST = starters; NST = non-starters.

Figures 4–6 show the differences between mesocycles for TM, TS and ACWR calculated through the s-RPE, TD and HSR across the in-season for the whole team.

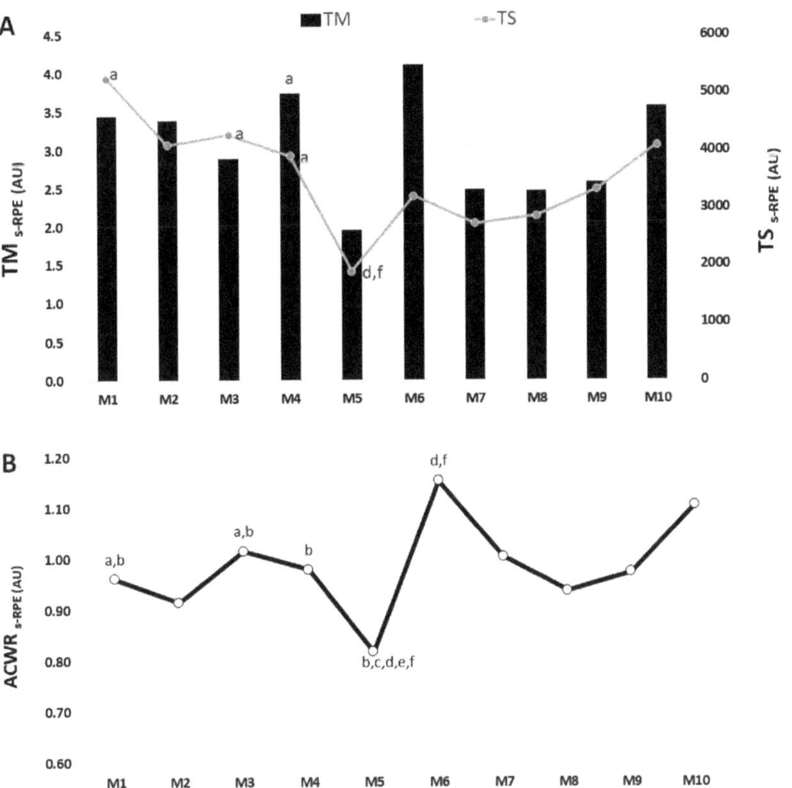

Figure 4. TM, TS (**A**) and ACWR (**B**) variations calculated through the s-RPE across 10 mesocycles for the whole team. M: mesocycle; a: difference from M5; b: difference from M6; c: difference from M7; d: difference from M8; e: difference from M9; f: difference from M10.

Overall, Figure 4A shows that the highest $TM_{s\text{-}RPE}$ occurred in M6 and the lowest value occurred in M5. There only was one significant difference for $TM_{s\text{-}RPE}$ in M4 > M5 (ES = 0.17). The highest $TS_{s\text{-}RPE}$ occurred in M1 and the lowest value occurred in M5. There was a significant difference in M1 > M5 (ES = 1.50); M3 > M5 (ES = 1.57); M4 > M5 (ES = 1.42); M5 < M8 (ES = −0.62) and <M10 (ES = −0.97).

Figure 4B shows that the highest $ACWR_{s\text{-}RPE}$ occurred in M6 while the lowest $ACWR_{s\text{-}RPE}$ occurred in M5. There were significant differences in M1 > M5 (ES = 1.63) and <M6 (ES = 7.60); M3 > M5 (ES = 11.75) and <M6 (ES = −10.69); M4 < M6 (ES = −1.42); M5 < M6 (ES = −8.75), <M7 (ES = −9.35), <M8 (ES = −9.25), <M9 (ES = −8.33) and <M10 (ES = −7.17); M6 > M8 (ES = −7.25) and >M10 (ES = 5.85).

Overall, Figure 5A shows that the highest TM_{TD} occurred in M6 and the lowest value in M1. There were significant differences in M1 < M2 (ES = −7.80), <M3 (ES = −5.70), <M4 (ES = −6.18), <M5 (ES = −3.81), <M6 (ES = −1.55) and <M7 (ES = −8.03); M2 < M4 (ES = −6.42); M4 > M7 (ES = −4.89) and M9 (ES = −0.93). The highest TS_{TD} occurred in M6 and the lowest value occurred in M2. There were significant differences in M1 < M2 (ES = −6.52), <M2 (ES = −5.35), <M3 (ES = −5.03) and <M10 (ES = −4.33); M2 < M4 (ES = −4.73), >M5 (ES = −2.92), >M7 (ES = −1.69); M3 > M5 (ES = −2.63), >M7 (ES = −1.63),

>M9 (ES = −3.00) and M4 > M5 (ES = −1.98), >M7 (ES = −1.51), >M9 (ES = −2.00). Additionally, M7 < M10 (ES = −3.52).

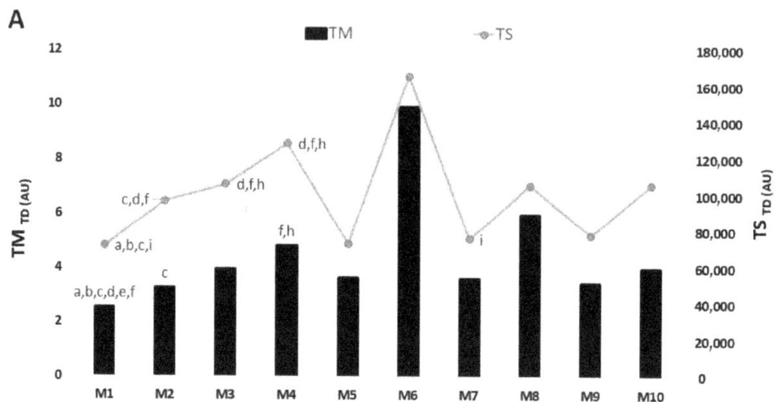

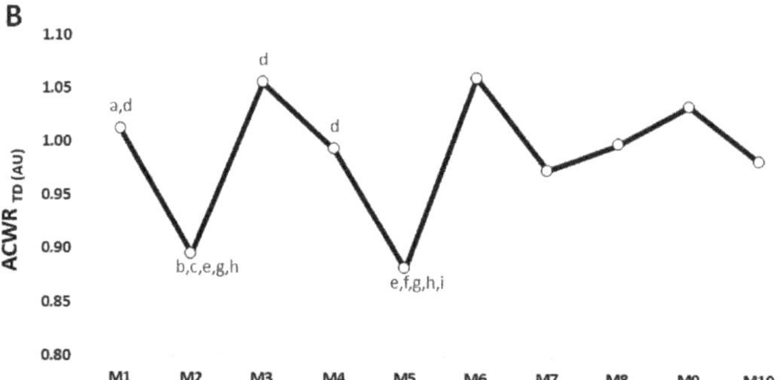

Figure 5. TM, TS (**A**) and ACWR (**B**) variations calculated through the total distance across 10 mesocycles for the whole team. M: mesocycle; a: difference from M2; b: difference from M3; c: difference from M4; d: difference from M5; e: difference from M6; f: difference from M7; g: difference from M8; h: difference from M9; i: difference from M10.

Figure 5B shows that the highest $ACWR_{TD}$ value occurred in M6, while the lowest value occurred in M5. There were significant differences in M1 > M2 (ES = −12.21) and >M5 (ES = −17.02). M2 < M3 (ES = −12.18), <M4 (ES = −12.05), <M6 (ES = −10.95), <M7 (ES = −13.75) and <M8 (ES = −13.42). M3 > M5 (ES = −12.99). M4 > M5 (ES = −15.64). M5 < M6 (ES = −14.30), <M7 (ES = −16.41), <M8 (ES = −25.59), <M9 (ES = −23.62) and <M10 (ES = −13.89).

Overall, Figure 6A showed that the highest TM_{HSR} occurred in M6 and the lowest value in M1. There were significant differences in M1 < M3 (ES = −5.42), < M6 (ES = −5.47). M2 < M6 (ES = −4.95). The highest TS_{HSR} occurred in M1 and the lowest value occurred in M5. There were significant differences in M2 > M6 (ES = 1.55), > M5 (ES = 0.16), > M7 (ES = 0.15), > M8 (ES = 0.32) and > M9 (ES = 0.40). M3 > M5 (ES = 0.15) and > M8 (ES = 0.19). M5 < M10 (ES = −0.79). Additionally, M7 < M10 (ES = −1.56).

In Figure 6B, the highest $ACWR_{HSR}$ value occurred in M6 while the lowest value occurred in M5. There were significant differences in M3 > M5 (ES = −5.05). M5 < M6 (ES = −4.93), < M8 (ES = −5.75), < M9 (ES = −5.78) and < M10 (ES = −5.21).

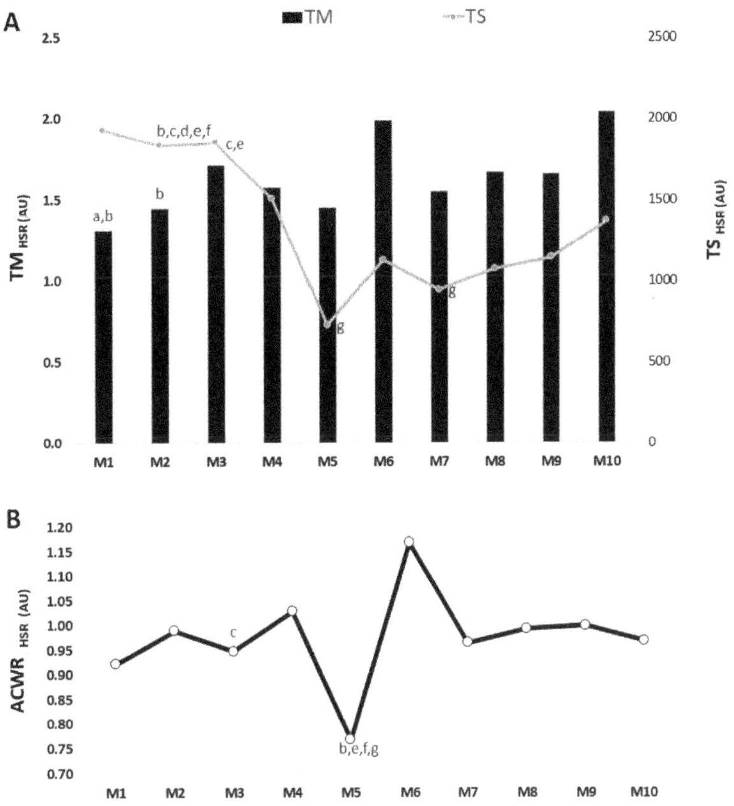

Figure 6. TM, TS (**A**) and ACWR (**B**) variations calculated through the HSR across 10 mesocycles for the whole team. M: mesocycle; a: difference from M3; b: difference from M6; c: difference from M5; d: difference from M7; e: difference from M8; f: difference from M9; g: difference from M10.

4. Discussion

The main purpose of the current study was to provide a description regarding training monotony (TM), strain (TS) and acute/chronic workload ratio (ACWR) based on perceived exertion, total distance (TD) and high-speed running (HSR) measures collected across in-season soccer. A secondary goal was to compare the time-related behavior of such metrics among starter and non-starter players. Our results in an elite European soccer team squad showed the following: (i) in the mesocycle with a greater number of matches disputed, higher values of various indices occurred, including either monotony (s-RPE, TD and HSR), strain (i.e., only TD in this specific case) or ACWR (s-RPE, TD and HSR); (ii) for all parameters considered, there were no significant differences between starters and non-starters; (iii) despite the similarities observed, players with a distinct status showed peak or lower values in distinct moments of the monitored period for some markers (e.g., lower $TM_{S\text{-}RPE}$ in the start and middle of the season, respectively, for non-starters and starters); (iv) higher monotony of perceived exertion was reported in the beginning, while for strain, it happened at the end of the season, independent of player status. In the following paragraphs, we will discuss the role of possible increase in match congestion on the data presented here while also accounting for the absence of differences between starters and non-starters and the common pattern of time-related variations in training monitoring parameters.

According to our results, the most intense period of games notably induced increases in monotony and ACWR (all variables) and concerning TS_{TD}. Eight soccer fixtures were played across one month, rendering an average of at least two per week during the mesocycle. Indeed, this can characterize a full, congested schedule as per previous definitions [22,23]. Despite the monotony of s-RPE being more than twofold above the suggested threshold of 2 AU [10] as in the case of starters, the total duration of training sessions decreased for such players, while the same was not valid for non-starters. This is possible given the requirements of the latter to be more involved in active training/matches during that moment of the season and given the likely need to rotate players. In fact, congested fixture periods are linked to the possibility of inducing greater TS [8], whilst they can impair physical match performance [23–25] and raise injury risk [22,26]. However, the values for ACWR were all below 1.3, independent of player status (see Table 2), which, in theory, may not represent exacerbated injury likelihood [27], despite the fact that such a question lacks consensus to date (see, for example, Impellizzeri et al. [28]). Based on these assumptions, it seems that adjustments promoted during training may help avoid worst scenarios relating to management of players' workloads across the most congested period of matches in a season. However, particular attention should be paid to non-starters since they presented a high monotony and no reduction in total training time as compared to previous ones and aligned with a partly higher strain (i.e., TS_{TD}) during the intense period of games.

One key finding of the present study was that when players were grouped according to their playing status as starters or non-starters, no significant differences were detected. This can suggest that contemporary soccer training methods require players to respond to stimuli delivered in a homogenous way, i.e., irrespective of whether generally starting the games or not. Importantly, one previous work verified opposing results, considering training TM and TS from accelerometer-derived variables, where starters showed greater values compared to their non-starter peers [14]. In contrast, non-starters may experience greater overall in-game physical exertion as compared to starters or players who participated in a whole match [29], and a similar condition was verified considering the most demanding passages of play [30]. Reports both have [31] and have not [32] confirmed the match–training-load associations in soccer. Of note, although there was no statistical significance here in the comparisons depending on player status, starters and non-starters reached maximal and minimal values for various markers at distinct moments. This finding can be related to distinct demands placed over each player across the season owing to situational-induced variations [33] and their prominent non-linear usage. Taken together, these assumptions could indicate that monitoring players on an individualized basis seems necessary, accounting for whether players generally start games on the pitch or the bench. Notwithstanding, traditional measures of workload such as TD and HSR may not be sensitive enough to detect possible status-related differences in monitoring strain, monotony and ACWR in training routines.

The findings from our investigation may assist in understanding the role of player status in various parameters used to control training in soccer as well potentially serving as a benchmark for future prescriptions and monitoring. Regardless of playing status and considering just the s-RPE, training monotony peaked at the beginning of the period (week 1), while the strain reached the largest values at the end (week 41). Such observations are different when compared to a six-week congestive period, which found lower values of monotony and strain in the first week, but the highest values in the last week for both variables [34]. The present results also disagree with the idea that a high degree of strain is often achieved when there is no competition (e.g., pre-season) [35]. High monotony early in the period may be indicative of either a poor ability of athletes to recognize the initial training loads or a true heavy stimulus applied, making it difficult to cope with as per the common fitness status of players at that moment. For example, the training session durations or strain levels were not the highest in the first week, whilst this does not hold true for TMs-RPE. Indeed, prior off-season training is recognized to impair physical capacity aspects [36] and it may have contributed to the prime TMs-RPE outputs.

Aside from the aforementioned potential derived implications (e.g., informing conditioning professionals on the effects of playing status and provision of reference values), a number of limitations of the present investigation should be highlighted. With the ever increasing energy requirements in soccer, the data gathered here may be outdated to some extent. The mostly descriptive nature of the work may limit its practical application. The generalizability of the results to other teams/countries, competitive standards and ages is also not warranted and requires replication studies. Complete description of training drills in further research can facilitate field implementation. Finally, co-variables such as match location, results and opponent quality should be considered in future studies as previously recommended [33].

5. Conclusions

To summarize, here, we observed across in-season soccer that spikes in training monotony, ACWR and strain for both internal and external load parameters (except regarding strain) may occur during match congestion intensification in elite soccer. Most importantly, apart from the extreme values being slightly discrepant (i.e., highest/lowest outcomes of the monitored markers varied according to playing status), starters and non-starters behaved equally across the period, thereby suggesting a lack of differences between them in the adjustments of training workloads during the period. Finally, the progression of the training cycle phases elicited distinct responses of monitoring indices, such as the monotony of perceived exertion, which reached peak values at the early season, and major strain was reported at the end-season stage. The results suggest that the training load and management of load were properly addressed, despite some play-time differences across the season. Moreover, the present study shows that it is possible to have a congested mesocycle with eight matches with higher workloads (M6). In addition, this is the first study to report data for the 10 mesocycles of the in-season period and could be considered a reference for future studies.

Author Contributions: Conceptualization, R.O. and F.M.C.; methodology, R.O. and F.M.C.; software, R.O., A.M. and M.N.; validation, R.O., A.M. and J.P.B.; formal analysis, R.O. and M.N.; investigation, R.O., L.H.P.V., A.M., J.P.B., M.N., B.M. and F.M.C.; resources, R.O., L.H.P.V., A.M., J.P.B., B.M. and F.M.C.; data curation, B.M.; writing—original draft preparation, R.O., L.H.P.V., A.M. and F.M.C.; writing—review and editing, R.O., L.H.P.V., A.M., J.P.B. and F.M.C.; visualization, R.O., L.H.P.V., A.M., J.P.B. and F.M.C.; supervision, R.O., A.M. and F.M.C.; project administration, R.O. and J.P.B.; funding acquisition, R.O. and J.P.B. All authors have read and agreed to the published version of the manuscript.

Funding: This research was funded by the Portuguese Foundation for Science and Technology, I.P., Grant/Award Number UIDP/04748/2020.

Institutional Review Board Statement: The study was conducted according to the guidelines of the Declaration of Helsinki and approved by the Ethics Committee of Polytechnic Institute of Santarém (252020Desporto).

Informed Consent Statement: Written informed consent was obtained from the participants to publish this paper.

Data Availability Statement: The data presented in this study are available on request from the corresponding author.

Acknowledgments: Luiz H. Palucci Vieira: ongoing PhD fellowship provided by São Paulo Research Foundation—FAPESP, under process number (2018/02965-7). "The opinions, hypotheses and conclusions or recommendations expressed in this material are the responsibility of the authors and do not necessarily reflect the views of FAPESP". Additionally, the authors would like to thank the team's coaches and players for their cooperation during all data collection procedures.

Conflicts of Interest: The authors declare no conflict of interest. The funders had no role in the design of the study; in the collection, analyses, or interpretation of data; in the writing of the manuscript, or in the decision to publish the results.

References

1. Hader, K.; Rumpf, M.C.; Hertzog, M.; Kilduff, L.P.; Girard, O.; Silva, J.R. Monitoring the Athlete Match Response: Can External Load Variables Predict Post-match Acute and Residual Fatigue in Soccer? A Systematic Review with Meta-analysis. *Sports Med.* **2019**, *5*, 48. [CrossRef]
2. Impellizzeri, F.M.; Marcora, S.M.; Coutts, A.J. Internal and External Training Load: 15 Years On. *Int. J. Sports Physiol. Perform.* **2019**, *14*, 270–273. [CrossRef]
3. Djaoui, L.; Haddad, M.; Chamari, K.; Dellal, A. Monitoring training load and fatigue in soccer players with physiological markers. *Physiol. Behav.* **2017**, *181*, 86–94. [CrossRef]
4. Rago, V.; Brito, J.; Figueiredo, P.; Costa, J.; Barreira, D.; Krustrup, P.; Rebelo, A. Methods to collect and interpret external training load using microtechnology incorporating GPS in professional football: A systematic review. *Res. Sports Med.* **2019**, *28*, 437–458. [CrossRef]
5. Boullosa, D.; Casado, A.; Claudino, J.G.; Jiménez-Reyes, P.; Ravé, G.; Castaño-Zambudio, A.; Lima-Alves, A.; de Oliveira, S.A.; Dupont, G.; Granacher, U.; et al. Do you Play or Do you Train? Insights from Individual Sports for Training Load and Injury Risk Management in Team Sports Based on Individualization. *Front. Physiol.* **2020**, *11*, 995. [CrossRef] [PubMed]
6. Moalla, W.; Fessi, M.S.; Farhat, F.; Nouira, S.; Wong, D.P.; Dupont, G. Relationship between daily training load and psychometric status of professional soccer players. *Res. Sports Med.* **2016**, *24*, 387–394. [CrossRef] [PubMed]
7. Manzi, V.; D'Ottavio, S.; Impellizzeri, F.M.; Chaouachi, A.; Chamari, K.; Castagna, C. Profile of Weekly Training Load in Elite Male Professional Basketball Players. *J. Strength Cond. Res.* **2010**, *24*, 1399–1406. [CrossRef]
8. Jaspers, A.; Brink, M.S.; Probst, S.G.M.; Frencken, W.G.P.; Helson, W.F. Relationships Between Training Load Indicators and Training Outcomes in Professional Soccer. *Sports Med.* **2017**, *47*, 533–544. [CrossRef] [PubMed]
9. Hulin, B.T.; Gabbett, T.J.; Lawson, D.W.; Caputi, P.; Sampson, J.A. The acute:chronic workload ratio predicts injury: High chronic workload may decrease injury risk in elite rugby league players. *Br. J. Sports Med.* **2016**, *50*, 231–236. [CrossRef]
10. Foster, C. Monitoring training in athletes with reference to overtraining syndrome. *Med. Sci. Sports Exerc.* **1998**, *30*, 1164–1168. [CrossRef]
11. Nobari, H.; Praça, G.M.; Clemente, F.M.; Pérez-Gómez, J.; Carlos Vivas, J.; Ahmadi, M. Comparisons of new body load and metabolic power average workload indices between starters and non-starters: A full-season study in professional soccer players. *Proceed. Inst. Mech. Eng. Part P J. Sports Eng. Technol.* **2021**, *235*, 105–113. [CrossRef]
12. Dalen, T.; Lorås, H. Monitoring Training and Match Physical Load in Junior Soccer Players: Starters versus Substitutes. *Sports* **2019**, *7*, 70. [CrossRef]
13. Nobari, H.; Oliveira, R.; Clemente, F.M.; Adsuar, J.C.; Pérez-Gómez, J.; Carlos-Vivas, J.; Brito, J.P. Comparisons of Accelerometer Variables Training Monotony and Strain of Starters and Non-Starters: A Full-Season Study in Professional Soccer Players. *Int. J. Environ. Res. Public Health* **2020**, *17*, 6547. [CrossRef]
14. Foster, C.; Florhaug, J.A.; Lori Gottschall, J.F.; Hrovatin, L.A.; Parker, S.; Doleshal, C.D.P. CBI-CDPBO1 and CBI-CDPBI1: CC-1065 analogs containing deep-seated modifications in the DNA binding subunit. *J. Strength Cond. Res.* **2001**, *15*, 109–115. [CrossRef]
15. Foster, C.; Hector, L.L.; Welsh, R.; Schrager, M.; Green, M.A.; Snyder, A.C. Effects of specific versus cross-training on running performance. *Eur. J. App. Physiol. Occup. Physiol.* **1995**, *70*, 367–372. [CrossRef] [PubMed]
16. Beato, M.; Devereux, G.; Stiff, A. Validity and reliability of global positioning system units (STATSports Viper) for measuring distance and peak speed in sports. *J. Strength Cond. Res.* **2018**, *32*, 2831–2837. [CrossRef]
17. Clemente, F.M.; Silva, R.; Castillo, D.; Los Arcos, A.; Mendes, B.; Afonso, J. Weekly Load Variations of Distance-Based Variables in Professional Soccer Players: A Full-Season Study. *Int. J. Environ. Res. Public Health* **2020**, *17*, 3300. [CrossRef] [PubMed]
18. Dalen-Lorentsen, T.; Bjørneboe, J.; Clarsen, B.; Vagle, M.; Fagerland, M.W.; Andersen, T.E. Does load management using the acute: Chronic workload ratio prevent health problems? A cluster randomised trial of 482 elite youth footballers of both sexes. *Br. J. Sports Med.* **2021**, *55*, 108–114. [CrossRef]
19. Impellizzeri, F.M.; Wookcock, S.; Coutts, A.J.; Fanchini, M.; McCall, A.; Vigotsky, A.D. Acute to random workload ratio is 'as' associated with injury as acute to actual chronic workload ratio: Time to dismiss ACWR and its components. *SportRxiv* **2020**. [CrossRef]
20. Myers, N.L.; Aguilar, K.V.; Mexicano, G.; Farnsworth, J.L.; Knudson, D.; Kibler, W.B.E.N. The Acute:Chronic Workload Ratio Is Associated with Injury in Junior Tennis Players. *Med. Sci. Sports Exerc.* **2020**, *52*, 1196–1200. [CrossRef]
21. Hopkins, W.G.; Marshall, S.W.; Batterham, A.M.; Hanin, J. Progressive statistics for studies in sports medicine and exercise science. *Med. Sci. Sports Exerc.* **2009**, *41*, 3–13. [CrossRef]
22. Dupont, G.; Nedelec, M.; McCall, A.; McCormack, D.; Berthoin, S.; Wisløff, U. Effect of 2 soccer matches in a week on physical performance and injury rate. *Am. J. Sports Med.* **2010**, *38*, 1752–1758. [CrossRef]
23. Lago-Peñas, C.; Rey, E.; Lago-Ballesteros, J.; Casáis, L.; Domínguez, E. The influence of a congested calendar on physical performance in elite soccer. *J. Strength Cond. Res.* **2011**, *25*, 2111–2117. [CrossRef] [PubMed]

24. Julian, R.; Page, R.M.; Harper, L.D. The Effect of Fixture Congestion on Performance During Professional Male Soccer Match-Play: A Systematic Critical Review with Meta-Analysis. *Sports Med.* **2021**, *51*, 255–273. [CrossRef] [PubMed]
25. Zanetti, V.; Carling, C.; Aoki, M.S.; Bradley, P.S.; Moreira, A. Are There Differences in Elite Youth Soccer Player Work Rate Profiles in Congested vs. Regular Match Schedules? *J. Strength Cond. Res.* **2021**, *35*, 473–480. [CrossRef]
26. Dellal, A.; Lago-Peñas, C.; Rey, E.; Chamari, K.; Orhant, E. The effects of a congested fixture period on physical performance, technical activity and injury rate during matches in a professional soccer team. *Br. J. Sports Med.* **2015**, *49*, 390–394. [CrossRef]
27. Gabbett, T.J. The training—injury prevention paradox: Should athletes be training smarter and harder? *Br. J. Sports Med.* **2016**, *50*, 273–280. [CrossRef]
28. Impellizzeri, F.M.; Woodcock, S.; Coutts, A.J.; Fanchini, M.; McCall, A.; Vigotsky, A.D. What Role Do Chronic Workloads Play in the Acute to Chronic Workload Ratio? Time to Dismiss ACWR and Its Underlying Theory. *Sports Med.* **2021**, *51*, 581–592. [CrossRef]
29. Gai, Y.; Volossovitch, A.; Leicht, A.S.; Gómez, M.-Á. Technical and physical performances of Chinese Super League soccer players differ according to their playing status and position. *Int. J. Perform. Anal. Sport* **2019**, *19*, 878–892. [CrossRef]
30. Fereday, K.; Hills, S.P.; Russell, M.; Smith, J.; Cunningham, D.J.; Shearer, D.; McNarry, M.; Kilduff, L.P. A comparison of rolling averages versus discrete time epochs for assessing the worst-case scenario locomotor demands of professional soccer match-play. *J. Sci. Med. Sport* **2020**, *23*, 764–769. [CrossRef]
31. Modric, T.; Versic, S.; Sekulic, D. Relations of the Weekly External Training Load Indicators and Running Performances in Professional Soccer Matches. *Sport Mont.* **2021**, *19*, 31–37.
32. Clemente, F.M.; Rabbani, A.; Conte, D.; Castillo, D.; Afonso, J.; Truman Clark, C.C.; Nikolaidis, P.T.; Rosemann, T.; Knechtle, B. Training/Match External Load Ratios in Professional Soccer Players: A Full-Season Study. *Int. J. Environ. Res. Public Health* **2019**, *16*, 3057. [CrossRef]
33. Gonçalves, L.G.C.; Kalva-Filho, C.A.; Nakamura, F.Y.; Rago, V.; Afonso, J.; Bedo, B.L.D.S.; Aquino, R. Effects of Match-Related Contextual Factors on Weekly Load Responses in Professional Brazilian Soccer Players. *Int. J. Environ. Res. Public Health* **2020**, *17*, 5163. [CrossRef]
34. Saidi, K.; Zouhal, H.; Rhibi, F.; Tijani, J.M.; Boullosa, D.; Chebbi, A.; Hackney, A.C.; Granacher, U.; Bideau, B.; Abderrahman, A.B. Effects of a six-week period of congested match play on plasma volume variations, hematological parameters, training workload and physical fitness in elite soccer players. *PLoS ONE* **2019**, *14*, e0219692. [CrossRef]
35. Haddad, M.; Stylianides, G.; Djaoui, L.; Dellal, A.; Chamari, K. Session-RPE Method for Training Load Monitoring: Validity, Ecological Usefulness, and Influencing Factors. *Front. Neurosci.* **2017**, *11*, 612. [CrossRef] [PubMed]
36. Silva, J.R.; Brito, J.; Akenhead, R.; Nassis, G.P. The transition period in soccer: A window of opportunity. *Sports Med.* **2016**, *46*, 305–313. [CrossRef]

Article

Relationships between Fitness Status and Match Running Performance in Adult Women Soccer Players: A Cohort Study

Lillian Gonçalves [1,*], Filipe Manuel Clemente [2,3], Joel Ignacio Barrera [4], Hugo Sarmento [4], Francisco Tomás González-Fernández [5], Luiz H. Palucci Vieira [6], António José Figueiredo [4], Cain C. T. Clark [7] and J. M. Cancela Carral [1]

1 Faculty of Educational Sciences and Sports Sciences, University of Vigo, 36005 Pontevedra, Spain; chemacc@uvigo.es
2 Escola Superior Desporto e Lazer, Instituto Politécnico de Viana do Castelo, Rua Escola Industrial e Comercial de Nun'Álvares, 4900-347 Viana do Castelo, Portugal; filipe.clemente5@gmail.com
3 Instituto de Telecomunicações, Delegação da Covilhã, 1049-001 Lisboa, Portugal
4 Research Unit for Sport and Physical Activity, Faculty of Sport Sciences and Physical Education, University of Coimbra, 3004-531 Coimbra, Portugal; jibarrera@outlook.es (J.I.B.); hg.sarmento@gmail.com (H.S.); afigueiredo@fcdef.uc.pt (A.J.F.)
5 Department of Physical Activity and Sport Sciences, Pontifical University of Comillas (Centro de Estudios Superiores Alberta Giménez), 07013 Palma, Spain; francis.gonzalez.fernandez@gmail.com
6 MOVI-LAB Human Movement Research Laboratory, School of Sciences, Graduate Program in Movement Sciences, Physical Education Department, UNESP São Paulo State University, Bauru 01140-070, Brazil; luiz.palucci@unesp.br
7 Centre for Intelligent Healthcare, Coventry University, Priory St, Coventry CV1 5FB, UK; cain.clark@coventry.ac.uk
* Correspondence: lilliangoncalves@ipvc.pt

Citation: Gonçalves, L.; Clemente, F.M.; Barrera, J.I.; Sarmento, H.; González-Fernández, F.T.; Palucci Vieira, L.H.; Figueiredo, A.J.; Clark, C.C.T.; Carral, J.M.C. Relationships between Fitness Status and Match Running Performance in Adult Women Soccer Players: A Cohort Study. *Medicina* **2021**, *57*, 617. https://doi.org/10.3390/medicina57060617

Academic Editors: Jan Bilski and Tatiana Moro

Received: 14 April 2021
Accepted: 11 June 2021
Published: 13 June 2021

Abstract: *Background and Objectives:* The aim of this study was twofold: (i) to analyze the relationships between fitness status (repeated-sprint ability (RSA), aerobic performance, vertical height jump, and hip adductor and abductor strength) and match running performance in adult women soccer players and (ii) to explain variations in standardized total distance, HSR, and sprinting distances based on players' fitness status. *Materials and Methods*: The study followed a cohort design. Twenty-two Portuguese women soccer players competing at the first-league level were monitored for 22 weeks. These players were tested three times during the cohort period. The measured parameters included isometric strength (hip adductor and abductor), vertical jump (squat and countermovement jump), linear sprint (10 and 30 m), change-of-direction (COD), repeated sprints (6 × 35 m), and intermittent endurance (Yo-Yo intermittent recovery test level 1). Data were also collected for several match running performance indicators (total distance covered and distance at different speed zones, accelerations/decelerations, maximum sprinting speed, and number of sprints) in 10 matches during the cohort. *Results*: Maximal linear sprint bouts presented large to very large correlations with explosive match-play actions (accelerations, decelerations, and sprint occurrences; $r = -0.80$ to -0.61). In addition, jump modalities and COD ability significantly predicted, respectively, in-game high-intensity accelerations ($r = 0.69$ to 0.75; $R^2 = 25\%$) and decelerations ($r = -0.78$ to -0.50; $R^2 = 23$–24%). Furthermore, COD had significant explanatory power related to match running performance variance regardless of whether the testing and match performance outcomes were computed a few or several days apart. *Conclusion*: The present investigation can help conditioning professionals working with senior women soccer players to prescribe effective fitness tests to improve their forecasts of locomotor performance.

Keywords: football; athletic performance; match analysis; sports training; GPS; high-intensity running

1. Introduction

Soccer matches represent a well-known intermittent mode of exercise in which short periods of intense efforts are interspaced by periods of low-to-moderate intensity [1,2]. Thus, players must maintain a desired level of running intensity and recover rapidly to perform to the best of their ability [3,4]. The literature has demonstrated that women soccer players may cover 9 to 11 km per match while spending 99 $\pm$ 8.3 m$\cdot$min^{-1} performing low-speed running and 9.7 $\pm$ 3.7 m$\cdot$min^{-1} performing high-speed running [2,5–8].

In female soccer, sprinting is considered a high-intensity effort [9], and high-speed activity is considered an essential component of matches. Usually, such efforts occur during decisive moments in a match [7], though they represent only 8% to 12% of the total distance covered in a typical match [10]. Additionally, female players were found to perform between 70 and 190 high-intensity runs (>19.8 km$\cdot$h^{-1}) during a match [5,10,11], covering between 210 and 520 m [6,7,12,13].

To sustain such efforts, female soccer players should present well-developed fitness statuses that allow them to meet the various demands of a match. Regarding sprinting performance, typical fitness status values observed in women soccer players suggest that they can cover 10 m in 2.31 $\pm$ 0.21 s and 25 m in 4.52 $\pm$ 0.20 s [14–17]. For another determinant variable (i.e., lower limb power), typical values exhibited by women soccer players are 30.1 $\pm$ 3.7 cm in the squat jump and 31.6 $\pm$ 4.0 cm in the countermovement jump [18]. Both sprinting and lower-limb power are neuromuscular determinants of soccer performance. However, the sport overwhelmingly involves running at low-to-moderate intensities—thus, good cardiorespiratory performance is crucial.

Female players usually present maximal oxygen uptake values between 49.4 and 56.7 mL/kg/min [2,17]. Based on one of the most common field-based tests used in soccer (namely, the Yo-Yo intermittent recovery test level 1), elite women soccer players can cover 1224 $\pm$ 255 m during the test, while players from lower divisions cover 826 $\pm$ 160 m [2,17].

Since high-intensity runs and sprinting tend to decrease at the end of the match, they could be associated with fatigue [19–22]. Therefore, sustaining good aerobic levels can help players avoid the effects of fatigue when performing power-related actions. Naturally, a player's performance will be affected by multiple factors, such as their position [23,24]. For instance, research indicates that central defenders perform fewer high-intensity runs than other players [23,24].

Fitness status can support match running performance—however, the strength of this relationship differs depending on the type of demand imposed on the player and the physical quality. For example, repeated sprint ability seems to be significantly correlated with total and high-intensity distances covered in matches [23–38]. Total distance also presented large correlations with high-intensity running activities and aerobic performance in field-based tests performed by male and female youth soccer players [29–42].

However, very few studies have tested the relationships between fitness status and match running performance among female soccer players. Nevertheless, it is pertinent to consider which kind of fitness status best relates to specific efforts in matches since match running performance is a determinant of a player's ability to sustain a high performance level. Understanding this matter will help to emphasize and specify the training process. However, fitness status changes over time. As such, analyzing the relationships between match running performance and fitness status in different moments throughout a season can help to explore whether these relationships are influenced by time.

Following the above discussion, the present study aims to (i) analyze the relationships between fitness status (repeated sprint ability (RSA), aerobic performance, vertical jump height, and anthropometry) and match running performance and (ii) run a regression analysis to explain variations in total distance, high-speed running (HSR), and sprinting distance. The hypothesis of the study is that match running variables are explained by the fitness status of the players.

2. Materials and Methods

2.1. Experimental Approach to the Study

This 22-week study followed an observational analytic cohort design. Players were assessed three times during the cohort (Figure 1). The first and second assessments were separated by four weeks, whereas the second and third assessments were separated by 18 weeks. The intervals were varied to determine the relationship between the physical capacities assessed with the match running and variations observed in total distance, HSR, and sprinting distance during matches.

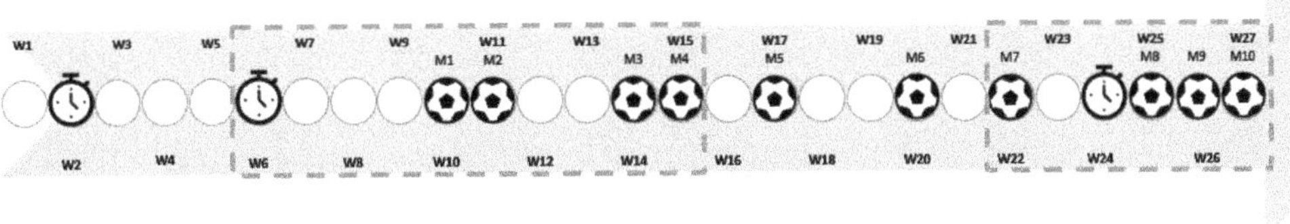

Figure 1. Timeline of the study.

Three participants were excluded from the analysis, and 22 participants remained. Assessment 2 was correlated with matches 1 to 4 (weeks 6 to 15), while assessment 3 was correlated with matches 7 to 10 (weeks 22 to 27). A correlation analysis was conducted between fitness variables and match running performance for each period of assessment. Additionally, multi-linear regression analysis was carried out considering the match running performance variables to determine how each of the three variables of interest influenced running performance.

2.2. Participants

Twenty-two women soccer players from a team participating in the first Portuguese league were observed (Table 1). The participants presented a mean age of 24.77 ± 6.49 years old and a height of 162.51 ± 7.08 cm. In the first assessment mean weight was 59.06 ± 9.50 kg. In the second assessment mean weight was 59.01 ± 9.30 kg and body mass 61.62 ± 9.50 kg. The sample included three goalkeepers, four external defenders, four central defenders, six midfielders, and five attackers. During the season, players participated in four training sessions per week and official matches on weekends.

The eligibility criteria that players had to meet to be included in the final sample were as follows: (i) completion of all three assessments; (ii) participation in at least 85% of training sessions, (iii) not being out of action for treatment for more than four weeks, and (iv) at least five years of experience.

Before the study began, all players were informed of the study's design and procedures. Afterward, each player signed an informed consent form. The study was approved by the local university (code: CTC-ESDL-CE001-2021; date: 18 March 2021) and followed the ethical standards as per the Declaration of Helsinki for studies involving humans.

Table 1. Physical fitness assessment (mean ± SD).

Measure	Women Soccer Players (n = 22)		
	Assessment 1	**Assessment 2**	**Assessment 3**
	Hip strength		
ADDs (kg)	-	34.66 ± 7.81	35.81 ± 7.22
ABDs (kg)	-	33.48 ± 5.87	34.40 ± 6.03
	Squat and countermovement jump		
SJ (cm)	25.33 ± 2.98	26.24 ± 3.09	23.85 ± 4.29
CMJ (cm)	27.26 ± 2.97	27.40 ± 3.51	24.17 ± 4.16
	Change-of-direction test		
COD (s)	5.73 ± 0.19	5.75 ± 0.18	5.80 ± 0.22
COD ($km{\cdot}h^{-1}$)	12.60 ± 0.40	12.53 ± 0.39	12.42 ± 0.46
COD ($m{\cdot}s^{-1}$)	3.50 ± 0.11	3.48 ± 0.10	3.45 ± 0.12
	Linear Sprinting		
10-m (s)	1.87 ± 0.08	1.90 ± 0.10	1.88 ± 0.10
10-m ($km{\cdot}h^{-1}$)	19.29 ± 0.84	18.98 ± 0.98	19.13 ± 0.99
10-m ($m{\cdot}s^{-1}$)	5.36 ± 0.23	5.27 ± 0.27	5.31 ± 0.27
30-m (s)	4.79 ± 0.22	4.77 ± 0.21	4.75 ± 0.23
30-m ($km{\cdot}h^{-1}$)	22.57 ± 1.05	22.64 ± 0.99	22.75 ± 1.05
30-m ($m{\cdot}s^{-1}$)	6.27 ± 0.29	6.29 ± 0.27	6.31 ± 0.29
	Repeated sprint ability test (RSA test)		
P_{max} (s)	380.81 ± 68.38	401.77 ± 74.47	448.63 ± 64.99
P_{min} (s)	240.44 ± 46.87	267.15 ± 46.29	295.53 ± 34.68
$P_{average}$ (s)	305.21 ± 48.93	321.83 ± 50.53	355.23 ± 39.31
FI (%)	4.61 ± 1.85	4.41 ± 1.65	5.02 ± 1.75
	Yo-Yo intermitteng recovery test- Level 1		
Stage (n)	14.62 ± 0.65	14.94 ± 0.77	15.15 ± 0.73
YYIR1, Distance (m)	677.78 ± 251.74	788.00 ± 219.89	682.66 ± 397.89
HR_{max} (bpm)	197.58 ± 5.33	197.50 ± 5.33	197.04 ± 5.25
$VO2_{max}$ ($mL{\cdot}kg^{-1}{\cdot}min^{-1}$)	41.79 ± 2.11	43.02 ± 1.85	43.56 ± 1.73

Note: $VO2_{max}$ was estimated by the next equation: Yo-Yo IR1 test: $VO2_{max}$ (mL/min/kg) = IR1 distance (m) × 0.0084 + 36.4 (Bangsbo. 2008); ADD: adductor strength; ABD: abductor strength; SJ: squat jump; CMJD: countermovement jump; COD: change-of-direction; P_{max}: maximum power at repeated-sprint test; P_{min}: minimum power at repeated-sprint test; $P_{average}$: average power at repeated-sprint test; FI: fatigue index at repeated-sprint test; YYIR1: intermittent recovery test level 1; HR_{max}: maximal heart rate; $VO2_{max}$: maximal oxygen uptake.

2.3. Measures

2.3.1. Physical Fitness Assessment

Between August and January, three fitness assessments with similar demands occurred in three microcycles. For each assessment period (week), three days were dedicated to run the tests, interspaced by 24 h between them. Players had 48 h of rest before the first day of assessments of each week analyzed.

In the first training session of the week, players were tested for anthropometry and hip adductor and abductor strength. In the second training session, vertical jump height, changes of direction, and linear speed were assessed. In the third session, the repeated sprint ability test and the Yo-Yo intermittent recovery test level 1 were applied.

These tests always occurred at the same time (7:30 p.m.) and location. The linear speed, repeated sprint ability, and Yo-Yo intermittent recovery tests were performed on synthetic turf without rain at a mean temperature of 19.5 ± 3.4 °C and a relative humidity of 63 ± 4%. A warm-up was performed before all evaluations. Warm-ups consisted of low and self-paced running, followed by calisthenic exercises in which players performed two sets of 10 repetitions of walking lunges, single-leg deadlifts, and fontal and lateral high knee movements.

Anthropometry

Body weight (kg) was measured without shoes with a bioelectrical impedance analysis (BIA) device (Tanita BC-730) to the nearest 0.1 kg. Height (cm) was measured using a stadiometer (Type SECA 225, Hamburg, Germany) to the nearest 0.1 cm.

Repeated Sprint Ability

The running anaerobic sprint test (RAST) test was applied to test players' repeated sprint abilities. This test consisted of six runs of 35 linear meters (each interspaced by 10 s of rest), with no COD required [43]. The time (sec) of each effort was recorded using a photocell timing gate (Photocells, Brower Timing System, UT, USA), with one device positioned at the starting line and the other positioned at the finish line. The device had a resolution of one-thousandth of a second. The minimum and maximum peak power and the fatigue index were determined using the following equation [43]: $\text{Power} = \frac{\text{Weight} \times \text{Distance}^2}{\text{Time}^3}$ and $\text{Fatigue} = \frac{\text{max}_{\text{power}} - \text{min}_{\text{power}}}{\text{Sum of 6 sprints (s)}}$.

Linear Sprinting

Players' 10- and 30-m linear sprint abilities were tested using photocell timing gates (Photocells, Brower Timing System, UT, USA) positioned at the start and finish lines. Participants began the test positioned 0.5 m behind the starting line in a two-point split stance. As with the repeated sprint test, the device used to measure the players' performance had a resolution of one-thousandth of a second. Each player's best result obtained from three separate trials was recorded as their sprint time.

Change-of-Direction

The zig-zag 20 m [40] test was used to assess COD. This test consists of four 5 m each set out at 100°. Times were once again recorded using photocells timing gates (Photocells, Brower Timing System, UT, USA) with a resolution of one-thousandth of a second. The typical error of the Photocells was between 0.04 and 0.06 s, while the smallest worthwhile change was between 0.11 and 0.17 s [41]. Subjects performed three trials, resting for at least three minutes between trials. The best time (lowest time in seconds) of the three trials was used for the analysis.

Squat and Countermovement Jump

Squat and countermovement jump heights were assessed, with the highest jumps (cm) recorded and used in the analysis. Both jumps were tested with an optical measurement system consisting of a transmitting and receiving bar (Optojump, Microgate, Bolzano, Italia).

Each participant started the squat jump test in a squat position (although self-selected, the recommendation was to stay approximately at 90° relative knee joint angle) with their hands on their waist. After spending three seconds in the squat position, the participant jumped by extending their legs and then landed in the same place. Each participant performed three trials, with 30 s of rest provided between jumps.

Each participant started the countermovement jump test from a standing position, with their hands on their waist. After spending three seconds in the standing position, the participant flexed their legs and then immediately extended them while jumping. Each participant performed three trials, with 30 s rest provided between jumps.

Yo-Yo Intermittent Recovery Test—Level 1

For the Yo-Yo Intermittent Recovery test, participants were to run 20 m from one mark to another and then return to the starting mark. After every 40 m covered, a 10-s recovery period is provided, during which time participants jog between two marks that are five meters apart (an audio beep is utilized to control participants' speed). The speed starts at 10 km/h, increasing progressively thereafter. The test ends when the athlete achieves voluntary exhaustion or does not reach one of the 20-m marks before or at the same time as the beep. At the end of the test, the number of completed levels and shuttles, as well as the total distance covered, were recorded. The total distance (meters) was recorded.

Hip Adductor and Abductor Strength

A dynamometer (Smart Groin Trainer, Neuro excellence, Portugal) was positioned on the thigh area of participants, who were asked to squeeze the tool for 20 s. Three trials were performed, with 10 s of rest between trials. The strength of the hip adductor and abductor was measured in kilograms. The highest value was used in the analysis.

2.3.2. Match Running Performance

During the match, participants used a Global Position System (GPS) (SPI HPU, GP-Sports, Canberra, Australia). This device has a frequency of 15 Hz and accelerometer of 100 Hz, 16 G Tri-axis, and a magnetometer of 50 Hz. Participants were asked to use a tight-fitting vest during the match and the device was placed between the left and right scapula. The GPS device collected the speed (km·h^{-1}), the maximal speed (km·h^{-1}), the number of sprints, the time of each sprint (sec), and accelerations and decelerations executed during each match observed. Speed achieved during a match was divided into the following 6 zones: zone 1 (0–5.9 km·h^{-1}), zone 2 (6–11.9 km·h^{-1}), zone 3 (12–13.9 km·h^{-1}), zone 4 (14–17.9 km·h^{-1}), zone 5 (18–23.9 km/h), and zone 6 (>24 km·h^{-1}). The acceleration and deceleration were also recorded and split into 3 zones: ace1 (1.0–1.9 m·s^2), ace2 (2.0–2.9 m·s^2), ace3 (3.0–4.0 m·s^2) and des1 (1.0–1.9 m·s^2), des2 (2.0–2.9 m·s^2), des3 (3.0–4.0 m·s^2). The external load collected for analysis were: total distance covered (m), the distance covered (m) in the different speed zones, accelerations (m·s^2), decelerations (m·s^2), the maximum speed achieved (km/h), and the number of sprints (n).

2.4. *Statistical Analysis*

Descriptive statistics were represented as mean ± SD. Normal distribution and homogeneity was tested with the Kolmogorov-Smirnov test on all data before analysis. A Pearson correlation coefficient *r* was used to examine the relationship between values of fitness assessment (hip strength (ADDs and ABDs); squat and countermovement jump (SJ and CMJ); change-of-direction test (COD in seconds); linear Sprinting (10 m and 30 m in seconds); repeated sprint ability test (P_{max}, P_{min} and FI); Yo-Yo intermittent recovery test 1 (YYIR1 distance)) and match running performance (total distance covered (D); speed achieved in zone 1 (Z1), zone 2 (Z2), zone 3 (Z3), zone 4 (Z4), zone 5 (Z5), and zone 6 (Z6); acceleration (ace1, ace2, ace3) and deceleration (des1, des2, des3); maximum speed achieved (MSA); and number of sprint (NS)). To interpret the magnitude of these correlations we adopted the following criteria: $r \leq 0.1$, trivial; $0.1 < r \leq 0.3$, small; $0.3 < r \leq 0.5$, moderate; $0.5 < r \leq 0.7$, large; $0.7 < r \leq 0.9$, very large; and $r > 0.9$, almost perfect [44]. The changes over the assessment were determined using repeated measures ANOVA. Significant main effects were subsequently analyzed using a Bonferroni post hoc test. Effect size is indicated with partial eta squared for Fs. To interpret the magnitude of the eta squared we adopted the following criteria: $\eta^2 = 0.02$, small; $\eta^2 = 0.06$, medium; and $\eta^2 = 0.14$ large. Regression analysis was used to identify which fitness outcomes can better explain match running performance. All variables were examined separately in this regression analysis. The magnitude of R2 was interpreted as follows: >0.02, small; >0.13, medium; >0.23, large. Data were analyzed using Statistica software (version 10.0; Statsoft, Inc., Tulsa, OK, USA).

3. Results

Descriptive statistics were calculated for each variable (see Tables 1 and 2 for more information).

A repeated measures ANOVA with participants' mean hip strength (ADDs and ABD) did not reveal any effect of assessment $F > 1$, in both cases. Another repeated measures ANOVA with participants' mean squat and countermovement jump (SJ and CMJ) did not reveal an effect of assessment in SJ, F (1.12) = 2.42, $p = 0.11$, $\eta^2 = 0.16$. However, data showed a significant effect of assessment in CMJ, F (1.12) = 6.13, $p = 0.01$, $\eta^2 = 0.33$. Continuing with the same type of repeated measures ANOVA analysis with participant 's mean change-of-direction (COD (s), COD (km·h^{-1}), and COD (m·s^{-1})) did not reveal any effect of assessment $F > 1$. In the same line, another ANOVA analysis with participants mean linear sprinting (10 m (s), 10 m (km·h^{-1}), 10 (m·s^{-1}), 30 m (s), 30 m (km·h^{-1}), 30 (m·s^{-1})) did not reveal any effect of assessment $F > 1$. A repeated measures ANOVA with participants' mean repeated sprint ability test (P_{max} (s), P_{min} (s), $P_{average}$ (s) and FI (%)) revealed an effect of assessment for P_{max} (s), P_{min} (s), and $P_{average}$ (s), F (1.12) = 4.86, $p = 0.01$, $\eta^2 = 0.28$, F (1.12) = 8.84, $p = 0.001$, $\eta^2 = 0.42$, and F (1.12) = 6.23, $p = 0.01$, $\eta^2 = 0.34$, respectively. Nevertheless, there was no effect of assessment for FI (%), $F > 1$. Particularly remarkable, a repeated measures ANOVA with participants' mean Yo-Yo intermittent recovery test level 1 (stage (n), YYIR1, distance (m), HR_{max} (bpm), and $V02_{max}$ (mL·kg^{-1}·min^{-1})) revealed an effect of assessment for stage (n), YYIR1, distance (m), and $V02_{max}$ (mL·kg^{-1}·min^{-1}), F (1.8) = 7.40, $p = 0.001$, $\eta^2 = 0.48$, F (1.8) = 7.40, $p = 0.001$, $\eta^2 = 0.48$, F (1.8) = 7.40, $p = 0.01$, $\eta^2 = 0.42$, respectively. However, HR_{max} (bpm) data did not show any effect of assessment, $F > 1$.

Table 2. Descriptive table of match running performance (mean ± SD).

Measure	Match 1 to Match 4 (*n*)	Match 7 to Match 10 (*n*)	Match 1 to Match 4 (*n* per min)	Match 7 to Match 10 (*n* per min)
D (m)	8091.47 ± 391.16	8383.01 ± 622.60	101.46 ± 9.21	99.00 ± 16.09
Z1 (m)	3143.72 ± 176.67	3283.83 ± 268.36	39.12 ± 3.83	39.07 ± 7.11
Z2 (m)	3189.10 ± 312.05	3183.08 ± 269.28	40.13 ± 4.86	37.37 ± 5.63
Z3 (m)	740.84 ± 51.08	758.21 ± 93.40	9.35 ± 1.02	8.90 ± 1.52
Z4 (m)	721.66 ± 40.98	792.56 ± 92.23	9.14 ± 0.85	9.36 ± 1.72
Z5 (m)	273.78 ± 31.33	339.08 ± 51.93	3.44 ± 0.24	4.00 ± 0.90
Z6 (m)	19.64 ± 8.16	23.78 ± 9.91	0.24 ± 0.10	0.27 ± 0.12
ace1 (*n*)	172.31 ± 10.76	175.53 ± 21.11	2.16 ± 0.20	2.06 ± 0.35
ace2 (*n*)	40.68 ± 4.63	41.48 ± 6.56	0.52 ± 0.07	0.49 ± 0.11
ace3 (*n*)	2.69 ± 0.29	2.50 ± 0.47	0.03 ± 0.00	0.03 ± 0.01
des1 (*n*)	146.21 ± 9.68	146.37 ± 16.34	1.84 ± 0.17	1.71 ± 0.25
des2 (*n*)	44.24. ± 4.75	43.91 ± 7.91	0.56 ± 0.08	0.52 ± 0.13
des3 (*n*)	14.40 ± 1.80	15.36 ± 3.03	0.18 ± 0.12	0.18 ± 0.05
MSA (km·h^{-1})	24.00 ± 0.38	24.49 ± 1.35	24.00 ± 0.38	24.49 ± 1.35
NS (*n*)	14.74 ± 8.99	14.62 ± 0.65	0.19 ± 0.12	0.20 ± 0.04

Note: *n* per minute was calculated considering the time in match; NS: number of sprints; MSA: maximum speed achieved; zone 1 (Z1), zone 2 (Z2), zone 3 (Z3), zone 4 (Z4), zone 5 (Z5), and zone 6 (Z6); acceleration (ace1, ace2, ace3) and deceleration (des1, des2, des3); D: distance covered.

The effect of match running performance tested repeatedly (D, Z1, Z2, Z3, Z4, Z5, Z6, ace1, ace2, ace3, des1, des2, des3, MSA and NS = between match 1 to match 4 (n), match 7 to match 10 (n), match 1 to match 4 (n per min), match 7 to match 10 (n per min)) did not reveal any effect of assessment of any studied variable, $F > 1$, in all cases.

On the basis of data obtained, correlations analysis was performed in order to find the possible association between fitness assessment and match running. First, we performed analysis of assessment 2 and matches 1–4, and second, assessment 3 and matches 7–10. Consequently, the correlation between fitness assessment and match running (assessment 2 and matches 1–4) are summarized in Table 3. No significant correlations were found between fitness assessment and the next variables of match running (D, Z1, Z2, Z3, Z4, Z5, ace1, ace2, des1, des2, MSA, and NS). However, negative correlation was found between 30 m linear sprinting and ace3, $r = -0.52$, $p = 0.24$. Crucially, other negative correlations were found between COD and Z6 and ace3 and des3 ($r = -0.57$, $p = 0.024$; $r = -0.59$, $p = 0.011$; $r = -0.50$, $p = 0.034$, respectively).

Correlation analysis was performed in order to find possible association between fitness assessment and match running (assessment 3 and matches 7–10). All data are summarized in Table 4. No significant correlations were found between fitness assessment and the next variables of match running (D, Z2, Z3, ace1, and des1). Nevertheless, positive correlation was found between SJ and ace3, des2, des3, and NS ($r = 0.75$, $p = 0.007$; $r = 0.64$, $p = 0.035$; $r = 0.63$, $p = 0.035$, and $r = 0.70$, $p = 0.016$, respectively). Other positives correlations were found between CMJ and Z1, Z4, ace2, ace3, des2, des3, and NS ($r = 0.61$, $p = 0.048$; $r = 0.63$, $p = 0.040$; $r = 0.64$, $p = 0.036$, $r = 0.69$, $p = 0.019$, $r = 0.67$, $p = 0.022$, $r = 0.62$, $p = 0.039$, and $r = 0.70$, $p = 0.016$, respectively). Furthermore, negative correlations were encountered between 10 m and ace2, $r = -0.61$, $p = 0.047$; des2, $r = -0.61$, $p = 0.050$; and NS $r = -0.75$, $p = 0.008$. Negative correlations were encountered between 30 m and Z5, Z6, ace2, ace3, des2, des3, MSA, and NS ($r = -0.63$, $p = 0.039$; $r = -0.70$, $p = 0.016$; $r = -0.68$, $p = 0.021$, $r = -0.77$, $p = 0.006$, $r = -0.68$, $p = 0.022$, $r = -0.68$, $p = 0.022$, $r = -0.68$, $p = 0.023$, and $r = -0.80$, $p = 0.003$, respectively). In the same line, more negative correlations were found between COD and Z4, Z5, Z6, ace2, ace3, des2, des3, and NS ($r = -0.68$, $p = 0.022$; $r = -0.80$, $p = 0.003$; $r = -0.77$, $p = 0.006$, $r = -0.76$, $p = 0.007$, $r = -0.84$, $p = 0.001$, $r = -0.78$, $p = 0.005$, $r = -0.75$, $p = 0.007$, and $r = -0.74$, $p = 0.010$, respectively). In addition, another positive correlation was encountered between FI and MSA, $r = 0.61$, $p = 0.043$.

Lastly, a multilinear regression analysis was performed to verify which variable of fitness assessment (agreement with the correlation analysis) could be used to better explain match running performance (See Table 5. for more information).

Table 3. Correlations between fitness assessment and match running (assessment 2 and matches 1–4).

Measure	D	Z1	Z2	Z3	Z4	Z5	Z6	ace1	ace2	ace3	Des1	Des2	Des3	MSA	NS
ADD (kg)	0.19	0.20	0.14	0.15	0.26	0.31	0.19	0.18	0.36	0.17	0.20	0.20	0.23	0.24	0.33
ABD (kg)	0.26	0.37	0.21	0.07	0.15	0.14	0.07	0.17	0.20	−0.06	0.19	0.16	0.10	0.29	0.11
SJ (cm)	−0.03	−0.01	−0.07	−0.04	0.03	0.11	0.21	−0.06	0.21	0.35	−0.01	−0.04	0.07	−0.01	0.25
CMJ (cm)	0.07	0.13	0.04	−0.02	−0.02	0.03	0.11	0.01	0.20	0.24	0.06	0.00	0.01	0.07	0.24
10-m (s)	0.15	0.13	0.21	0.14	0.07	−0.11	−0.11	0.17	−0.13	−0.34	0.16	0.10	−0.17	0.07	−0.29
30-m (s)	0.17	0.13	0.26	0.19	0.07	−0.24	−0.40	0.26	−0.15	−0.53 *	0.25	0.14	−0.25	0.09	−0.27
COD (s)	0.02	0.11	0.09	−0.10	−0.25	−0.46	−0.57 *	0.02	−0.25	−0.59 *	0.05	−0.12	−0.50 *	−0.04	−0.17
P_{max} (W)	0.00	0.08	−0.07	−0.13	0.01	0.14	0.21	−0.12	0.11	0.22	−0.08	−0.08	0.09	0.06	0.19
P_{min} (W)	0.26	0.30	0.21	0.17	0.26	0.23	0.20	0.18	0.33	0.09	0.22	0.20	0.13	0.25	0.27
FI (%)	−0.22	−0.13	−0.27	−0.32	−0.20	0.00	0.13	−0.32	−0.12	0.22	−0.29	−0.27	0.02	−0.13	0.03
YYIR1 (m)	0.12	−0.09	0.20	0.30	0.28	0.15	0.05	0.19	0.20	0.05	0.19	0.22	0.14	0.10	0.19

ADD: adductor strength; ABD: abductor strength; SJ: squat jump; CMJD: countermovement jump; COD: change-of-direction; P_{max}: maximum power at repeated-sprint test; P_{min}: minimum power at repeated-sprint test; $P_{average}$: average power at repeated-sprint test; FI: fatigue index at repeated-sprint test; YYIR1: intermittent recovery test level 1; HR_{max}: maximal heart rate; $VO2_{max}$: maximal oxygen uptake; NS: number of sprints; MSA: maximum speed achieved; zone 1 (Z1), zone 2 (Z2), zone 3 (Z3), zone 4 (Z4), zone 5 (Z5), and zone 6 (Z6); acceleration (ace1, ace2, ace3) and deceleration (des1, des2, des3); D: distance covered; *: significant at $p < 0.05$.

Table 4. Correlations between fitness status and match running (assessment 3 and matches 7 to 10).

Measure	D	Z1	Z2	Z3	Z4	Z5	Z6	ace1	ace2	ace3	Des1	Des2	Des3	MSA	NS
ADD (kg)	−0.02	−0.05	0.05	0.11	−0.07	−0.31	−0.50	0.07	−0.22	−0.22	0.12	−0.07	−0.37	−0.23	−0.15
ABD (kg)	−0.10	−0.13	−0.03	0.01	−0.15	−0.32	−0.41	−0.07	−0.28	−0.17	−0.03	−0.16	−0.28	−0.27	−0.19
SJ (cm)	0.46	0.48	0.34	0.45	0.53	0.57	0.56	0.44	0.59	0.76 *	0.42	0.64 *	0.63 *	0.46	0.70 *
CMJ (cm)	0.60	0.61 *	0.50	0.60	0.63 *	0.60	0.48	0.57	0.64 *	0.69 *	0.56	0.68 *	0.63 *	0.56	0.70 *
10-m (s)	−0.52	−0.51	−0.47	−0.52	−0.52	−0.47	−0.43	−0.52	−0.61 *	−0.55	−0.49	−0.60 *	−0.56	−0.55	−0.75 *
30-m (s)	−0.57	−0.59	−0.47	−0.48	−0.57	−0.63 *	−0.70 *	−0.53	−0.68 *	−0.77 *	−0.51	−0.68 *	−0.68 *	−0.67 *	−0.80 *
COD (s)	−0.49	−0.51	−0.29	−0.45	−0.68 *	−0.80 *	−0.77 *	−0.50	−0.76 *	−0.84 *	−0.45	−0.78 *	−0.75 *	−0.56	−0.74 *
P_{max} (W)	0.43	0.52	0.39	0.21	0.23	0.30	0.36	0.35	0.30	0.51	0.30	0.41	0.31	0.51	0.49
P_{min} (W)	0.12	0.17	0.11	−0.01	0.02	0.12	−0.01	0.08	0.02	0.16	0.05	0.14	0.07	0.04	0.35
FI (%)	0.47	0.56	0.43	0.27	0.28	0.31	0.46	0.39	0.36	0.54	0.35	0.43	0.34	0.62 *	0.42
YYIR1 (m)	−0.41	−0.47	−0.42	−0.28	−0.19	−0.08	−0.25	−0.36	−0.19	−0.30	−0.35	−0.27	−0.12	−0.56	−0.08

ADD: adductor strength; ABD: abductor strength; SJ: squat jump; CMJD: countermovement jump; COD: change-of-direction; P_{max}: maximum power at repeated-sprint test; P_{min}: minimum power at repeated-sprint test; $P_{average}$: average power at repeated-sprint test; FI: fatigue index at repeated-sprint test; YYIR1: intermittent recovery test level 1; HR_{max}: maximal heart rate; $VO2_{max}$: maximal oxygen uptake; NS: number of sprints; MSA: maximum speed achieved; zone 1 (Z1), zone 2 (Z2), zone 3 (Z3), zone 4 (Z4), zone 5 (Z5), and zone 6 (Z6); acceleration (ace1, ace2, ace3) and deceleration (des1, des2, des3); D: distance covered; *: significant at $p < 0.05$.

Table 5. Values of regression analysis explaining fitness assessment and match running performance.

Measure	Measure	Assessment 2 and Matches 1–4					
		b *	SE of b *	R^2	Adjusted R^2	F	p
30-m (s)	ace3 (n)	−0.39	0.21	0.15	0.10	3.29	0.08
COD (s)	Z6 (m)	−0.55	0.19	0.30	0.25	7.71	0.01 *
	ace3 (n)	−0.56	0.19	0.31	0.27	8.36	0.009 *
	des3 (n)	−0.49	0.20	0.24	0.19	5.71	0.027 *
		Assessment 3 and matches 7–10					
SJ (cm)	ace3 (n)	0.50	0.22	0.25	0.20	5.04	0.04 *
	des2 (n)	0.27	0.24	0.77	0.01	1.25	0.28
	des3 (n)	0.36	0.24	0.13	0.07	2.25	0.15
	NS (n)	0.43	0.23	0.18	0.13	3.45	0.08
CMJ (cm)	Z1 (m)	0.33	0.24	0.11	0.05	1.93	0.18
	Z4 (m)	0.39	0.23	0.15	0.10	2.78	0.11
	ace2 (n)	0.35	0.24	0.12	0.07	2.14	0.16
	ace3 (n)	0.50	0.22	0.25	0.20	4.98	0.04 *
	des2 (n)	0.34	0.24	0.11	0.05	1.93	0.18
	des3 (n)	0.40	0.23	0.15	0.10	2.79	0.11
	NS (n)	0.45	0.23	0.45	0.15	3.95	0.06
10-m (s)	ace2 (n)	−0.19	0.24	0.37	-	0.61	0.44
	des2 (n)	−0.12	0.25	0.16	-	0.26	0.61
	NS (n)	−0.33	0.23	0.11	0.06	2.04	0.17
30-m (s)	Z5 (m)	−0.25	0.24	0.01	0.01	1.13	0.03
	Z6 (m)	−0.10	0.24	0.01	-	0.17	0.68
	Ace2 (n)	−0.26	0.24	0.07	0.01	1.18	0.29
	Ace3 (n)	−0.18	0.25	0.03	-	0.56	0.46
	Des2 (n)	−0.20	0.25	0.03	-	0.65	0.42
	Des3 (n)	−0.23	0.24	0.05	-	0.88	0.36
	MSA (km/h)	−0.39	0.23	0.12	0.09	3.11	0.19
	NS (n)	−0.41	0.22	0.16	0.11	3.24	0.09
COD (s)	Z4 (m)	−0.49	0.22	0.24	0.19	4.84	0.04 *
	Z5 (m)	−0.57	0.21	0.31	0.27	7.04	0.01 *
	Z6 (m)	−0.33	0.24	0.10	0.05	1.83	0.19
	Ace2 (n)	−0.55	0.22	0.30	0.25	6.53	0.02 *
	Ace3 (n)	−0.40	0.23	0.16	0.10	2.92	0.10
	Des2 (n)	−0.54	0.22	0.23	0.24	6.31	0.02 *
	Des3 (n)	−0.48	0.23	0.23	0.18	4.45	0.052
	NS (n)	−0.59	0.21	0.35	0.30	7.98	0.01 *
FI (%)	MSA (km/h)	−0.08	0.25	0.01	-	0.10	0.75

SJ: squat jump; CMJD: countermovement jump; COD: change-of-direction; P_{max}: maximum power at repeated-sprint test; P_{min}: minimum power at repeated-sprint test; $P_{average}$: average power at repeated-sprint test; FI: fatigue index at repeated-sprint test; YYIR1: intermittent recovery test level 1; HR_{max}: maximal heart rate; $VO2_{max}$: maximal oxygen uptake; NS: number of sprints; MSA: maximum speed achieved; zone 1 (Z1), zone 2 (Z2), zone 3 (Z3), zone 4 (Z4), zone 5 (Z5) and zone 6 (Z6); acceleration (ace1, ace2, ace3) and deceleration (des1, des2, des3); D: distance covered; *: significant at $p < 0.05$.

4. Discussion

The main aim of the current study was to determine the magnitude of relationships between various fitness status measures (strength, power, single/repeated sprinting, and intermittent endurance) and match running performance in adult women soccer players competing at a high level. We also aimed to explain the match running variations based on fitness status. The main findings in the present Portuguese players indicate the following: (1) correlations between fitness and match running performance were dependent on the time frame separating the testing battery and the collection of running performance during actual match-play. (2) With only rare exceptions, isolated strength, intermittent endurance, and repeated sprint ability performance were not associated with, nor did they predict, match running performance. (3) Even considering the fact that tests for separate maximal sprint bouts (10 and 30 m) were largely to very largely associated, they failed to significantly explain the variance of match-play (e.g., explosive) locomotor variables. (4) Jump and

COD ability clearly allowed a good (medium-to-large) prediction to be obtained of in-game high-intensity accelerations and decelerations, respectively. Finally, (5) the latter evaluation method was the only fitness indicator that had significant power to predict match running performance independent of the interval between testing and match performance.

Some innovative aspects in the current study should be highlighted. First, while the majority of previous works adopted only correlations as a statistical treatment to evaluate the possible link between fitness status and match running performance in soccer [41,45–48], a decision was made to move further when providing recommendations mainly based on regression analysis, which reveals the weighted influence of players' physical capacity on their locomotor outputs during match-play. In addition, various investigations on the subject have tested players at a single time point [36–38,40,41,49,50]. Meanwhile, here, two distinct approaches were considered using a cohort design, one testing associations between fitness status determined near the match occurrences and another with a longer interval between them. Most importantly, less than 3% of evidence on the complexity of fitness-match running performance relationships in a soccer context [51] were derived from scientific studies including female players according to knowledge collated in reviews [27,52]. Nonetheless, only intermittent endurance (Yo-Yo IR1/IE2), aerobic fitness (laboratory treadmill tests) [2,53], and Wingate measurements [11] were previously related to match running performance in women's soccer. Again, this reinforces the originality of data presented in the current work and supports the critical appraisal of the findings' strengths and weaknesses, which is developed in the following paragraphs.

An important finding of the present investigation is that a fitness testing battery seems to have a relatively short expiration date to help preview match physical performance in female soccer players. Such is indicated by the frequency and strength of correlations between fitness status measures and match running performance, as well as the number of variables involved, which varied in the distinct moments. According to a recent critique piece [53–57], manipulating the interval between players' evaluations and matches was never previously addressed when the objective was to understand their associations. Here, when looking at fitness assessment 2, we noticed only four moderate-to-large (ranging from −0.59 to −0.50) correlations with matches 1 to 4 (e.g., COD and 30 m sprint tests with in-game very-high intensity accelerations) (Table 3). All these were performed across a 10-week period, where fitness tests and matches were separated by three to nine weeks.

In contrast, more than 30 large to very large correlation coefficients (ranging from −0.84 to 0.75) were found between fitness assessment 3 parameters and running outputs during matches 7–10 (e.g., SJ, CMJ, 10/30 m sprint and COD with in-game sprint occurrences) (Table 4). Such second analysis comprised a 6-week period in total, with an interval between test and match equal to no more than three weeks. Naturally, changes occurring between fitness assessments were not explained in this work, but possibly may be affected by the training process [26,58,59]. Reports have not yet confirmed the same for match running performance, even though these are related to each other crosswise [3,15,57]. However, based on the current observations, the usefulness of some fitness data may become outdated (or at least its relevance might be reduced) after approximately a month in women's soccer.

The local strength of hip adductor and abductor muscles and intermittent high-intensity or endurance running bouts were not associated, nor were they predictors. However, 10- to 30-m sprint performances were largely to very largely related to match running outputs in the female players of the present study, though their shared variance had no statistical significance.

When comparing such results to those presented in the available literature, discrepancies are identified. For example, this was the case in intermittent endurance capacity, which was previously linked to match running performance in female senior players [3,57] as well as in senior and youth male populations [27]. However, recent studies have demonstrated that running outputs during small-sided games were associated with the outputs obtained in competitions [54] independent of the player's intermittent endurance profile [55]. Force, maximal velocity, and aerobic and anaerobic resistance are important fitness components arguably contributing to sustaining physical efforts experienced in soccer match-play, thereby representing frequent determinants to winning [56,57].

Notwithstanding, many issues likely compromise the utility of some fitness assessment protocols in the current format. For example, testing maximal sprint ability (single or successive running bouts) using only linear paths is criticized nowadays given the curved trajectory of most explosive in-game actions [58]. The very short intervals often offered between repeated all-out efforts do not match those encountered in actual matches [59]. Occurrences of near-to-maximal displacements can also be very uncommon in elite standards [60]. Hip adduction/abduction strength allows one to discriminate between distinct performance levels [61], yet the effective contribution of hip muscle strength to running kinematics is low [62,63]. To summarize, such points of view are in alignment with our results. The external validity of some popular fitness status markers in women's soccer is not always supported, and its indiscriminate use needs to be re-thought.

Conditioning professionals need to be aware of assessment tools that can remain consistent over time when classifying players based on their fitness status, as well as the potential implications in terms of match performance. In this sense, although the construct validity of the various methods tested here has been challenged, jump and COD ability provided reasonable predictions (R^2 ranging from 23 to 25%, respectively) (Table 5) for in-game high-intensity accelerations and decelerations. Furthermore, COD predicted match running performance regardless time between testing and match observations. In other words, regression models with inputs being COD data remained significant from assessment 2/matches 1–4 until assessment 3/matches 7–10.

Studies that have aimed to extract the most relevant game indicators in soccer suggest that accelerations and decelerations are among the main components of athletes' external loads [64]. Interestingly, decelerations are more frequent than accelerations in soccer match-play [65]. In addition to being paramount to COD performance, skilled decelerations are also fundamental to a range of match events (e.g., rapid changes in speed, cutting maneuvers, and regaining ball possession) [66]. Therefore, change-of-direction seems to be a sensitive indicator of fitness status in female soccer players, as it may provide meaningful information about the next match profile, in particular the players' deceleration performance.

Aside from the novelty of the current investigation, a number of limitations are recognized and need to be accounted for in future work, as well as when making interpretations and generalizations based on the present evidence. For one, female players were grouped regardless of the general exertion of their positional role during matches. Studies in male soccer players have shown that fitness status and match running performance relationships can be position dependent [25,26,53]. Another limitation is that the recommended sample size (N = 80 players) as per Gregson et al. [67] was not met in the present investigation, though this has frequently been the case in similar research. It is possible given the practical difficulty of involving multiple clubs that the pertinence of large/potentially heterogeneous datasets in solving this problem also lacks consensus. In addition, none of the conducted tests required players to rely on technical-tactical performance. Instead, they evaluated physical capacity markers. Adopting protocols that more closely mimic game demands could enhance the ecological validity and, in turn, the predictive ability of a testing battery in informing to some extent and advancing physical performance during matches [54]. Finally, the games involved different opposition over the course of the study, which may have impacted the running demands completed in matches.

5. Conclusions

Including a change-of-direction ability test seems pertinent when assessing women soccer players, as it may partly predict match-play running performance regardless of whether the time separating the assessment and competition is shorter (testing immediately prior to/after competing) or longer (test-match moments interspaced by at least three weeks). It is something that provides preliminary evidence about the construct validity of COD testing and its likely robustness regarding common time-related changes in match running performance.

We also demonstrated that the further apart a fitness testing battery is carried out in relation to actual matches, the lower its value in predicting in-game running outputs. Finally, caution is required concerning conditioning professionals' extensive use of common testing procedures, such as isolated maximal sprints, intermittent high-intensity actions, or endurance bouts, as evidence in the present study revealed that these do not always provide useful information for forecasting inter-individual variations in match-play locomotor performance.

Author Contributions: L.G., F.M.C., and J.M.C.C. led the project, established the protocol and wrote and revised the original manuscript. J.I.B. and H.S. collected the data and wrote and revised the original manuscript. F.T.G.-F., C.C.T.C., A.J.F., and L.H.P.V. wrote and revised the original manuscript. All authors have read and agreed to the published version of the manuscript.

Funding: This research received no external funding.

Institutional Review Board Statement: The study was conducted according to the guidelines of the Declaration of Helsinki and approved by the Institutional Review Board (or Ethics Committee) of the Polytechnic Institute of Viana do Castelo. School of Sport and Leisure (code: CTC-ESDL-CE001-2021).

Informed Consent Statement: Informed consent was obtained from all subjects involved in the study.

Acknowledgments: Filipe Manuel Clemente: This work is funded by Fundação para a Ciência e Tecnologia/Ministério da Ciência. Tecnologia e Ensino Superior through national funds and, when applicable, co-funded via EU funds under the project UIDB/50008/2020. Hugo Sarmento gratefully acknowledges the support of a Spanish government subproject "Integration ways between qualitative and quantitative data, multiple case development, and synthesis review as the main axis for an innovative future in physical activity and sports research" [PGC2018-098742-B-C31] (Ministerio de Economía y Competitividad. Programa Estatal de Generación de Conocimiento y Fortalecimiento Científico y Tecnológico del Sistema I+D+i), which is part of the coordinated project "New approach of research in physical activity and sport from mixed methods perspective" (NARPAS_MM) [SPGC201800X098742CV0]. Luiz H Palucci Vieira: ongoing PhD fellowship from São Paulo Research Foundation–FAPESP, under process number [#2018/02965-7].

Conflicts of Interest: The authors declare no conflict of interest.

References

1. Manson, S.; Brughelli, M.; Harris, N. Physiological Characteristics of International Female Soccer Players. *J. Stregth Cond. Res.* **2014**, *28*, 308–318. [CrossRef] [PubMed]
2. Krustrup, P.; Mohr, M.; Ellinsgaard, H.; Bangsbo, J. Physical Demands during an Elite Female Soccer Game: Importance of Training Status. *Med. Sci. Sports Exerc.* **2005**, *37*, 1242–1248. [CrossRef] [PubMed]
3. Milanović, Z.; Sporiš, G.; James, N.; Trajković, N.; Ignjatović, A.; Sarmento, H.; Trecroci, A.; Mendes, B. Physiological Demands, Morphological Characteristics, Physical Abilities and Injuries of Female Soccer Players. *J. Hum. Kinet.* **2017**, *60*, 77–83. [CrossRef] [PubMed]
4. Davis, J.; Brewer, J. Applied Physiology of Female Soccer Players. *Sports Med.* **1993**, *16*, 180–189. [CrossRef] [PubMed]
5. Stepinski, M.; Ceylan, H.I.; Zwierko, T. Seasonal Variation of Speed, Agility and Power Performance in Elite Female Soccer Players: Effect of Functional Fitness. *Phys. Activ. Rev.* **2020**, *8*, 16–25. [CrossRef]
6. Datson, N.; Hulton, A.; Andersson, H.; Lewis, T.; Weston, M.; Drust, B.; Gregson, W. Applied Physiology of Female Soccer: An Update. *Sports Med.* **2015**, *44*, 1225–1240. [CrossRef]
7. Datson, N.; Drust, B.; Weston, M.; Jarman, I.; Lisboa, P.; Gregson, W. Match Physical Performance of Elite Female Soccer Player during International Competition. *J. Stregth Cond. Res.* **2017**, *31*, 2379–2387. [CrossRef]
8. Trewin, J.; Meylan, C.; Varley, M.C.; Cronin, J. The Match-to-Match Variation of Match-Running in Elite Female Soccer. *J. Sci. Med. Sport* **2018**, *21*, 196–201. [CrossRef]
9. Vescovi, J.D. Sprint Profile of Professional Female Soccer Players during Competitive Matches: Female Athletes in Motion (FAiM) study. *J. Sports Sci.* **2012**, *30*, 37–41. [CrossRef]
10. Haugen, T.; Tønnessen, E.; Seiler, S. Speed and Countermovement-Jump Characteristics of Elite Female Soccer Players, 1995–2010. *Int. J. Sports Physiol. Perform.* **2012**, *7*, 340–349. [CrossRef]

11. McCormack, W.; Stout, J.; Wells, A.; Gonzalez, A.; Mangine, G.; Hoffman, J. Predictors of High-Intensity Running Capacity in Collegiate Women during a Soccer Game. *J. Stregth Cond. Res.* **2014**, *28*, 964–970. [CrossRef] [PubMed]
12. Andersson, H.; Randers, M.; Heiner-Moller, A.; Krustrup, P.; Mohr, M. Elite Female Soccer Players Perform More High-Intensity Running When Playing in International Games Compared with Domestic League Games. *J. Stregth Cond. Res.* **2010**, *24*, 912–919. [CrossRef]
13. Niessen, M.; Hartmann, U.; Martı, V. ScienceDirect Women's Football: Player Characteristics and Demands of the Game. *J. Sport Health Sci.* **2014**, *3*, 258–272.
14. Hoare, D.G.; Warr, C.R. Talent Identification and Women's Soccer: An Australian Experience. *J. Sport Sci.* **2010**, *18*, 751–758. [CrossRef] [PubMed]
15. Andersson, H.; Raastad, T.; Nilsson, J.; Paulsen, G.K.R.A.N.; Garthe, I.N.A.; Kadi, F. Neuromuscular Fatigue and Recovery in Elite Female Soccer: Effects of Active Recovery. *Med. Sci. Sports Exerc.* **2008**, *40*, 372–380. [CrossRef] [PubMed]
16. Sjökvist, J.; Laurent, M.; Richardson, M.; Curtner-Smith, M.; Holmber, H.-C.; Bishop, P. Recovery from High-intensity Training Sessions in Female Soccer Players. *J. Stregth Cond. Res.* **2014**, *25*, 1726–1735. [CrossRef] [PubMed]
17. Paulsen, K.; Butts, C.; McDermott, B. Observation of women soccer players' physiology during a single season. *J. Stregth Cond. Res.* **2018**, *32*, 1702–1707. [CrossRef] [PubMed]
18. Castagna, C.; Castellini, E. Vertical Jump Performance in Italian Male and Female National Team Soccer Players. *J. Strength Cond. Res.* **2013**, *27*, 1156–1161. [CrossRef]
19. Sedano, S.; Vaeyens, R.; Philippaerts, R.; Redondo, J.; Cuadrado, G. Anthropometric and Anerobic Fitness Profile of Elite and Non-Elite Female Soccer Players. *J. Sports Med. Phys. Fit.* **2009**, *49*, 387–394.
20. Krustrup, P.; Zebis, M.; Jensen, J.M.; Mohr, M. Game-Induced Fatigue Patterns in Elite Female Soccer. *J. Strength Cond. Res.* **2010**, *24*, 437–441. [CrossRef]
21. Mujika, I.; Santisteban, J.; Impellizzeri, F.M.; Castagna, C. Fitness Determinants of Success in Men's and Women's Football. *J. Sport Sci.* **2009**, *27*, 107–114. [CrossRef] [PubMed]
22. Mara, J.; Thompson, K.; Pumpa, K.L.; Ball, N. Periodisation and Physical Performance in Elite Female Soccer Players. *Int. J. Sports Physiol. Perform.* **2015**, *10*, 664–669. [CrossRef]
23. Rampinini, E.; Coutts, A.J.; Castagna, C.; Sassi, R.; Impellizzeri, F.M. Variation in Top Level Soccer Match Performance. *Int. J. Sports Med.* **2007**, *28*, 1018–1024. [CrossRef] [PubMed]
24. Buchheit, M.; Mendez-Villanueva, A.; Quod, M.J.; Poulos, N.; Bourdon, P. Determinants of the Variability of Heart Rate Measures during a Competitive Period in Young Soccer Players. *Eur. J. Appl. Physiol.* **2010**, *109*, 869–878. [CrossRef]
25. Haddad, H.; Simpson, B.M.; Buchheit, M.; Di Salvo, V.; Mendez-Villanueva, A. Peak Match Speed and Maximal Sprinting Speed in Young Soccer Players: Effect of Age and Playing Position. *Int. J. Sports Physiol. Perform.* **2015**, *10*, 888–896. [CrossRef] [PubMed]
26. Buchheit, M.; Saint, P.; Football, G.; Mendez-villanueva, A.; Simpson, B.M.; Bourdon, P.C. Match Running Performance and Fitness in Youth Soccer. *Int. J. Sports Med.* **2010**, *11*, 818–825. [CrossRef]
27. Aquino, R.; Carling, C.; Maia, J.; Vieira, L.H.P.; Wilson, R.S.; Smith, N.; Almeida, R.; Gonçalves, L.G.C.; Kalva-Filho, C.A.; Garganta, J.; et al. Relationships between Running Demands in Soccer Match-Play, Anthropometric, and Physical Fitness Characteristics: A Systematic Review. *Int. J. Perform. Anal. Sport* **2020**, *20*, 534–555. [CrossRef]
28. Stolen, T.; Chamari, K.; Castagna, C.; Wisloff, U. Physiology of Soccer. *Sports Med.* **2005**, *3*, 50–60.
29. Castellano, J.; Alvarez-Pastor, D.; Bradley, P.S. Evaluation of Research Using Computerised Tracking Systems (Amisco and Prozone) to Analyse Physical Performance in Elite Soccer: A Systematic Review. *Sports Med. Auckl. N. Z.* **2014**, *44*, 701–712. [CrossRef]
30. Carling, C.; Nelson, L. The Role of Motion Analysis in Elite Soccer: Contemporary Performance Measurements Techniques and Work Rate Data. *Sports Med.* **2008**, *38*, 839–862. [CrossRef]
31. Carling, C. Interpreting Physical Performance in Professional Soccer Match-Play: Should We Be More Pragmatic in Our Approach? *Sports Med. Auckl. N. Z.* **2013**, *43*, 655–663. [CrossRef]
32. Bangsbo, J.; Mohr, M.; Krustrup, P. Physical and Metabolic Demands of Training and Match-Play in the Elite Foot- Ball Player. *J. Sports Sci.* **2006**, 665–674. [CrossRef]
33. Castagna, C.; Impellizzeri, F.M.; Manzi, V.; Ditroilo, M. The Assessment of Maximal Aerobic Power with the Multistage Fitness Test in Young Women Soccer Players. *J. Strength Cond. Res.* **2010**, 1488–1494. [CrossRef] [PubMed]
34. Krustrup, P.; Mohr, M.; Amstrup, T.; Rysgaard, T.; Johansen, J.; Steensberg, A.; Pedersen, P.K.; Bangsbo, J. The Yo-Yo Intermittent Recovery Test: Physiological Response, Reliability, and Validity. *Med. Sci. Sports Exerc.* **2003**, *35*, 697–705. [CrossRef] [PubMed]
35. Fernandes-da-Silva, J.; Castagna, C.; Teixeira, A.S.; Carminatti, L.J.; Guilherme, L.; Guglielmo, A. The Peak Velocity Derived from the Carminatti Test Is Related to Physical Match Performance in Young Soccer Players The Peak Velocity Derived from the Carminatti Test Is Related to Physical Match. *J. Sports Sci.* **2016**, *34*, 2238–2245. [CrossRef]
36. Aquino, R.; Vieira, L.H.P.; Oliveira, L.D.P.; Gonçalves, L.G.C.; Santiago, P.R.P. Relationship between Field Tests and Match Running Performance in High-Level Young Brazilian Soccer Players. *J. Sports Med. Phys. Fit.* **2018**, *58*, 256–262.
37. Castagna, C.; Impellizzeri, F.; Cecchini, E.M.C.; Rampinini, E.; Alvarez, J. Effects of Intermittent-Endurance Fitness on Match Performance in Young Male Soccer Players. *J. Stregth Cond. Res.* **2009**, *23*, 1954–1959. [CrossRef]
38. Castagna, C.; Manzi, V.; Impellizzeri, F.; Weston, M.; Barbero Alvarez, J.C. Relationship Between Endurance Field Tests and Match Performance in Young Soccer Players. *J. Strength Cond. Res.* **2010**, *24*, 3227–3233. [CrossRef] [PubMed]

39. Rago, V.; Silva, J.R.; Mohr, M.; Barreira, D.; Krustrup, P.; Rago, V.; Silva, J.R.; Mohr, M.; Barreira, D.; Krustrup, P. The Inter-Individual Relationship between Training Status and Activity Pattern during Small-Sided and Full-Sized Games in Professional Male Football Players. *Sci. Med. Footb.* **2018**, *2*, 115–122. [CrossRef]
40. Rebelo, A.; Brito, J.; Seabra, A.; Oliveira, J.; Krustrup, P. Physical Match Performance of Youth Football Players in Relation to Physical Capacity. *Eur. J. Sport Sci.* **2014**, *14*, S148–S156. [CrossRef] [PubMed]
41. Abt, G.; Lovell, R.I.C. The Use of Individualized Speed and Intensity Thresholds for Determining the Distance Run at High-Intensity in Professional Soccer. *J. Sports Sci.* **2009**, *27*, 893–898. [CrossRef] [PubMed]
42. Mohr, M.; Krustrup, P.; Bangsbo, J. Match Performance of High-Standard Soccer Players with Special Reference to Development of Fatigue. *J. Sports Sci.* **2003**, *21*, 519–528. [CrossRef] [PubMed]
43. Cipryan, L.; Gajda, V. The Influence of Aerobic Power on Repeated Anaerobic Exercise in Junior Soccer Players. *J. Hum. Kinet.* **2011**, *28*, 63–71. [CrossRef]
44. Granier, P.; Mercier, B.; Mercier, J.; Anselme, F.; Préfaut, C. Aerobic and Anaerobic Contribution to Wingate Test Performance in Sprint and Middle-Distance Runners. *Eur. J. Appl. Physiol.* **1995**, *70*, 58–65. [CrossRef]
45. Metaxas, T.I. Match Running Performance of Elite Soccer Players: Vo2max and Players Position Influences. *J. Strength Cond. Res.* **2021**, *35*, 162–168. [CrossRef]
46. Redkva, P.E.; Paes, M.R.; Fernandez, R.; da-Silva, S.G. Correlation between Match Performance and Field Tests in Professional Soccer Players. *J. Hum. Kinet.* **2018**, *62*, 213–219. [CrossRef]
47. Aslan, A.; Acikada, C.; Güvenç, A.; Gören, H.; Hazir, T.; Özkara, A. Metabolic Demands of Match Performance in Young Soccer Players. *J. Sports Sci. Med.* **2012**, *11*, 170–179. [PubMed]
48. Carling, C.; Le Gall, F.; McCall, A.; Nedelec, M.; Dupont, G. Are Aerobic Fitness and Repeated Sprint Ability Linked to Fatigue in Professional Soccer Match-Play? A Pilot Study. *J. Athl. Enhanc.* **2013**, 2. [CrossRef]
49. Bangsbo, J.; Lindquist, F. Comparison of Various Exercise Tests with Endurance Performance during Soccer in Professional Players. *Int. J. Sports Med.* **1992**, *13*, 125–132. [CrossRef] [PubMed]
50. Palucci Vieira, L.H.; Arins, F.B.; Guglielmo, L.G.A.; de Lucas, R.D.; Carminatti, L.J.; Santiago, P.R.P. Game Running Performance and Fitness in Women's Futsal. *Int. J. Sports Med.* **2021**, *42*, 74–81.
51. Mendez-Villanueva, A.; Buchheit, M. Physical Capacity–Match Physical Performance Relationships in Soccer: Simply, More Complex. *Eur. J. Appl. Physiol.* **2011**, *111*, 2387–2389. [CrossRef] [PubMed]
52. Paul, D.J.; Nassis, G.P. Physical Fitness Testing in Youth Soccer: Issues and Considerations Regarding Reliability, Validity and Sensitivity. *Pediatr. Exerc. Sci.* **2015**, *27*, 301–313. [CrossRef]
53. Bradley, P.S.; Mohr, M.; Bendiksen, M.; Randers, M.B.; Flindt, M.; Barnes, C.; Hood, P.; Gomez, A.; Andersen, J.L.; Di Mascio, M.; et al. Sub-Maximal and Maximal Yo–Yo Intermittent Endurance Test Level 2: Heart Rate Response, Reproducibility and Application to Elite Soccer. *Eur. J. Appl. Physiol.* **2011**, *111*, 969–978. [CrossRef] [PubMed]
54. Aquino, R.; Melli-Neto, B.; Ferrari, J.V.S.; Bedo, B.L.S.; Vieira, L.H.P.; Santiago, P.R.P.; Gonçalves, L.G.C.; Oliveira, L.P.; Puggina, E.F. Validity and Reliability of a 6-a-Side Small-Sided Game as an Indicator of Match-Related Physical Performance in Elite Youth Brazilian Soccer Players. *J. Sports Sci.* **2019**, *37*, 2639–2644. [CrossRef] [PubMed]
55. Stevens, T.G.A.; Ruiter, C.J.D.; Beek, P.J.; Savelsbergh, G.J.P. Validity and Reliability of 6-a-Side Small-Sided Game Locomotor Performance in Assessing Physical Fitness in Football Players. *J. Sports Sci.* **2016**, *34*, 527–534. [CrossRef] [PubMed]
56. Aquino, R.; Munhoz Martins, G.H.; Palucci Vieira, L.H.; Menezes, R.P. Influence of Match Location, Quality of Opponents, and Match Status on Movement Patterns in Brazilian Professional Football Players. *J. Strength Cond. Res.* **2017**, *31*, 2155–2161. [CrossRef]
57. Granero-Gil, P.; Bastida-Castillo, A.; Rojas-Valverde, D.; Gómez-Carmona, C.D.; de la Cruz Sánchez, E.; Pino-Ortega, J. Influence of Contextual Variables in the Changes of Direction and Centripetal Force Generated during an Elite-Level Soccer Team Season. *Int. J. Environ. Res. Public Health* **2020**, *17*, 967. [CrossRef]
58. Fitzpatrick, J.F.; Linsley, A.; Musham, C. Running the Curve: A Preliminary Investigation into Curved Sprinting during Football Match-Play. *Sport Perform. Sci. Rep.* **2019**, *55*, v1.
59. Schimpchen, J.; Skorski, S.; Nopp, S.; Meyer, T. Are "Classical" Tests of Repeated-Sprint Ability in Football Externally Valid? A New Approach to Determine in-Game Sprinting Behaviour in Elite Football Players. *J. Sports Sci.* **2016**, *34*, 519–526. [CrossRef] [PubMed]
60. Buchheit, M.; Simpson, B.M.; Hader, K.; Lacome, M. Occurrences of Near-to-Maximal Speed-Running Bouts in Elite Soccer: Insights for Training Prescription and Injury Mitigation. *Sci. Med. Footb.* **2020**, 1–6. [CrossRef]
61. Prendergast, N.; Hopper, D.; Finucane, M.; Grisbrook, T.L. Hip Adduction and Abduction Strength Profiles in Elite, Sub-Elite and Amateur Australian Footballers. *J. Sci. Med. Sport* **2016**, *19*, 766–770. [CrossRef]
62. Hannigan, J.J.; Osternig, L.R.; Chou, L.-S. Sex-Specific Relationships Between Hip Strength and Hip, Pelvis, and Trunk Kinematics in Healthy Runners. *J. Appl. Biomech.* **2018**, *34*, 76–81. [CrossRef] [PubMed]
63. Brund, R.B.K.; Rasmussen, S.; Nielsen, R.O.; Kersting, U.G.; Laessoe, U.; Voigt, M. The Association between Eccentric Hip Abduction Strength and Hip and Knee Angular Movements in Recreational Male Runners: An Explorative Study. *Scand. J. Med. Sci. Sports* **2018**, *28*, 473–478. [CrossRef]

64. Pino-Ortega, J.; Rojas-Valverde, D.; Gómez-Carmona, C.D.; Rico-González, M. Training Design, Performance Analysis, and Talent Identification—A Systematic Review about the Most Relevant Variables through the Principal Component Analysis in Soccer, Basketball, and Rugby. *Int. J. Environ. Res. Public Health* **2021**, *18*, 2642. [CrossRef]
65. Harper, D.J.; Carling, C.; Kiely, J. High-Intensity Acceleration and Deceleration Demands in Elite Team Sports Competitive Match Play: A Systematic Review and Meta-Analysis of Observational Studies. *Sports Med. Auckl. N. Z.* **2019**, *49*, 1923–1947. [CrossRef]
66. Mara, J.K.; Thompson, K.G.; Pumpa, K.L.; Morgan, S. The Acceleration and Deceleration Profiles of Elite Female Soccer Players during Competitive Matches. *J. Sci. Med. Sport* **2017**, *20*, 867–872. [CrossRef] [PubMed]
67. Gregson, W.; Drust, B.; Atkinson, G.; Salvo, V.D. Match-to-Match Variability of High-Speed Activities in Premier League Soccer. *Int. J. Sports Med.* **2010**, *31*, 237–242. [CrossRef] [PubMed]

 medicina

Article

Characteristics of Pedaling Muscle Stiffness among Cyclists of Different Performance Levels

Isaac López-Laval [1,*], Rafel Cirer-Sastre [2], Francisco Corbi [2] and Sebastian Sitko [1]

1 Faculty of Health and Sport Sciences (FCSD), University of Zaragoza, 22002 Huesca, Spain; sebastiansitko@yahoo.es

2 Institut Nacional d'Educació Física de Catalunya (INEFC), Universitat de Lleida (UdL), 25192 Lleida, Spain; rcirer@inefc.es (R.C.-S.); f@corbi.neoma.org (F.C.)

* Correspondence: isaac@unizar.es

Abstract: *Background and Objectives:* The aim of the present study was to compare the impact of an incremental exercise test on muscle stiffness in the rectus femoris (RF), vastus lateralis (VL), biceps femoris (BF), and gastrocnemius (GL) among road cyclists of three performance levels. *Materials and Methods:* The study group consisted of 35 cyclists grouped according to their performance level; elite ($n = 10$; professional license), sub-elite ($n = 12$; amateur license), and recreational ($n = 13$; cyclosportive license). Passive muscle stiffness was assessed using myometry before and after an incremental exercise test. *Results:* There was a significant correlation between time and category in the vastus lateralis with stiffness increases in the sub-elite ($p = 0.001$, Cohen's $d = 0.88$) and elite groups ($p = 0.003$, Cohen's $d = 0.72$), but not in the recreational group ($p = 0.085$). Stiffness increased over time in the knee extensors (RF, $p < 0.001$; VL, $p < 0.001$), but no changes were observed in the knee flexors (GL, $p = 0.63$, BF, $p = 0.052$). There were no baseline differences among the categories in any muscle. *Conclusions:* Although the performance level affected VL stiffness after an incremental exercise test, no differences in passive stiffness were observed among the main muscles implicated in pedaling in a resting state. Future research should assess whether this marker could be used to differentiate cyclists of varying fitness levels and its potential applicability for the monitoring of training load.

Keywords: cyclist; myometry; stiffness; incremental cycling test

Citation: López-Laval, I.; Cirer-Sastre, R.; Corbi, F.; Sitko, S. Characteristics of Pedaling Muscle Stiffness among Cyclists of Different Performance Levels. *Medicina* **2021**, 57, 606. https://doi.org/10.3390/medicina57060606

Academic Editor: Simon M. Fryer

Received: 20 May 2021
Accepted: 9 June 2021
Published: 11 June 2021

Publisher's Note: MDPI stays neutral with regard to jurisdictional claims in published maps and institutional affiliations.

1. Introduction

Road cycling is a popular endurance sport characterized by its cyclic nature, variable intensity, and large training volumes [1]. Among the key determinants of road cycling performance, maximal oxygen uptake (VO_{2max}) stands out as one of the values that best represents cardiorespiratory fitness [2]. Furthermore, this parameter has been proposed to classify endurance athletes based on fitness level [3]. Other performance determinants in road cycling include cycling economy or efficiency, tactical and technical skills, psychological resilience, body composition in accordance with the cycling discipline, and muscles' mechanical properties adapted to the riders' specialty [2,4]. The muscle mechanical properties might differ among cycling specialties: the most powerful riders are characterized by greater and shorter muscles and a greater predominance of fast twitch fibers when compared to climbers or time-trialists [5]. The relationship between cardiorespiratory fitness and muscle performance factors has been detailed by Hoper et al. (2013), who determined that the percentage of muscle fibers is influenced by the VO_{2max} of the cyclist [6].

Among the mechanical characteristics of the muscle, stiffness provides information regarding its intrinsic property and post-effort response; furthermore, it is one of the main parameters that characterizes the viscoelastic properties of the myofascial complex [7]. A proper conceptualization of muscle stiffness requires an analysis of both muscular architecture and its functional aspect. Concretely, it should be defined as the biomechanical capacity of the tissue (characterized by the type, number, and composition of its muscular fibers) that impedes stretching and distensions [8].

Current literature establishes different conclusions regarding stiffness levels [9–12]. On the one hand, optimal levels of musculotendinous stiffness are highly correlated to significant increases in muscle performance, especially in situations where the stretch-shorter cycle component is key to optimal performance [9–11]. On the other hand, higher stiffness can also be considered a potential threat, since a greater risk of injury has been reported in those athletes who presented greater levels of muscular stiffness as a consequence of high training loads [12]. The literature regarding endurance activities is scarce but has concluded that the reduction in musculotendinous stiffness observed in these types of sports is a consequence of the fatigue generated by submaximal muscle contractions that are sustained in time [13,14]. Related to this, García-Manso et al. (2011) determined that the loss in contractile capacity induced by a long-distance race reflects changes in the neuromuscular response and fluctuations in the contractile capacity of the muscle [13]. Furthermore, Andonian et al. (2016) reported a decrease in quadriceps stiffness caused by an extreme mountain ultra-marathon [14].

Since road cycling is an endurance sport with no impact and in which power is applied to the pedals, the lower limb muscle stiffness of cyclists might differ from other endurance disciplines. To the authors' knowledge, only four previous studies have analyzed road cycling muscle stiffness with non-invasive tools [15–18]. Several aspects could be highlighted from these studies: muscular stiffness seems to be an important contributory factor to sprint performance [15] and is proportional to the cyclists' power output during sprints [16]. Klich et al. (2020) observed higher stiffness in sprinters compared to endurance track cyclists [17]. In this same sense, it is important to highlight the results obtained by Ditroilo et al. (2011), who established that cyclists with higher baseline muscular stiffness suffered greater stiffness losses under fatigue than those with lower baseline levels [18].

The active and passive measurement of muscle stiffness is normally a complex procedure as it requires either muscular biopsies or repeated maximal isometric contractions [8]. Both methods may generate pain and require recovery after the procedure [10]. The utilization of non-invasive techniques for passive measurement of muscle stiffness could be highlighted as a viable alternative, especially when used in a field setting. Tensiomyography, elastography, electromyography, and ultrasounds have been the preferred non-invasive methods in recent years [10], although they still require educated staff and extended time periods for data obtention [8]. The MyotonPRO® (Myoton Ltd., Tallinn, Estonia) is a non-invasive tool that allows for the measurement of passive muscle stiffness through short oscillatory impulses [19] that are generated on the skin and over the area of the analyzed muscle [20]. Previous studies have demonstrated that the device is valid and reliable (ICC = 0.75–0.96; R^2 = 0.95) [8,21] and has been used to measure the main muscles that participate in the pedaling action: rectus femoris (RF), vastus lateralis (VL) [22], hamstring [23], and gastrocnemius (GL) [20]. Klich et al. (2019) proposed myotonometry as an easy and suitable tool to assess the viscoelastic characteristics of muscles in cyclists [24].

It is accepted that both fatigue and performance level of the athlete can influence the changes in muscle stiffness [15,18,20]. Given that previous studies have determined a relationship between stiffness and performance level in other sports, it could be hypothesized that this relationship could also exist in an endurance discipline such as road cycling. However, to date, the differences in muscle stiffness of road cyclists of different performance levels have not been examined. Further, the relationship between initial stiffness level and the response to fatigue after an incremental test is also an area of enquiry that has yet to be investigated. Accordingly, the aim of the present study was to compare the impact of an incremental exercise test on muscle stiffness among road cyclists of three performance levels.

2. Materials and Methods

2.1. Research Design

Thirty-five participants completed this cross-sectional study. The same order was followed during each individual assessment: anthropometry, evaluation of passive muscle stiffness (stiffness pre), incremental exercise test, and assessment of passive muscle stiffness after the incremental exercise test (stiffness post) (see Figure 1). In addition, during the assessment of passive muscle stiffness, the order was always the same between before and after the incremental exercise test and was standardized to avoid the influence of recovery time in muscle stiffness. Measurements were always taken in the same place (University Lab, Río Isuela Sport Center, Huesca, Spain), with a mean temperature of 21 ± 2 °C and mean relative humidity of 52% ± 9%). All participants were assessed during the preparatory phase of the annual training cycle, between the months of September and October and coinciding with the first preparatory meso-cycle. The evaluations were always carried out on Saturdays and Sundays at the same time (between 10 and 12 in the morning) to standardize the measurements and thus organize the schedule of travel to the laboratory. A 48-h rest period was established prior to the initial measurement to guarantee an adequate baseline assessment without fatigue. The assessment of the stiffness was carried out immediately after the completion of the incremental exercise test.

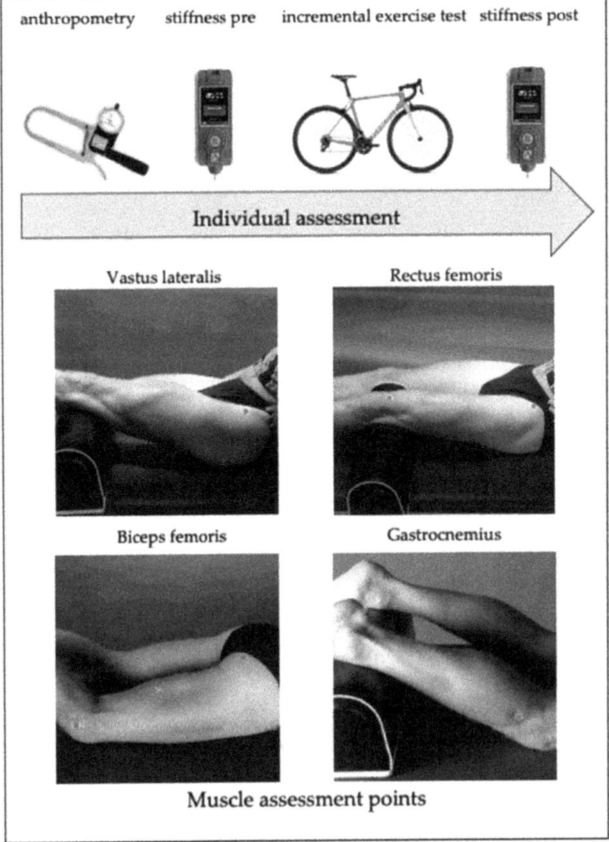

Figure 1. Experimental approach timeline and muscle assessment points. Blue marks: proximal and distal measurement area. Yellow mark: Myoton assessment point.

Assessment points placed at VL: 2/3 on the line from the anterior spina iliaca to the lateral side of the patella, RF: 50% on the line from the anterior spina iliaca superior to the superior part of the patella, BF: 50% on the line between the ischial tuberosity and the lateral epicondyle of the tibia, and GL:1/3 of the line between the head of the heel [25].

2.2. Participants

G∗Power version 3.1.9.2 [26] was used to estimate the required sample size in a 2 × 3 mixed design for a minimum expected effect size (Cohen's F) of 0.4, an α level of 0.05, and a power $(1-\beta)$ of 0.95. This procedure returned a minimum number of 30 participants [17]. Thirty-five male cyclists volunteered to participate in the present study. The main characteristics of the participants are reported in Table 1. Participants were allocated to groups according to their performance level: elite, cyclists with a professional license (n = 10); sub-elite, cyclists with an U23 or amateur license (n = 12); recreational, cyclists that participate in cyclosportive events (n = 13). The inclusion criteria were to own a professional, U23, cyclosportive, or masters license. The exclusion criteria were (a) surgical procedures and injuries in the six months prior to the study and (b) use of performance-enhancing drugs in the six months prior to the study. After being informed of the benefits and potential risks of the investigation, all participants signed an informed consent form. The study followed the ethical guidelines of the 2013 Declaration of Helsinki and received approval from the Research Ethics Committee of the autonomous region of Aragon, Spain (the approval code is PI19/447, approved on 4th December 2019).

Table 1. Summary of the characteristics of the participants.

	Recreational (n = 13)	Sub-Elite (n = 12)	Elite (n = 10)
Age (y)	42.3 ± 7.3	38.2 ± 10.4	32.3 ± 9.2 $^{<R\ (p = 0.037)}$
Height (cm)	177. 7 ± 8.8	179.25 ± 1.7	177.7 ± 8.3
Mass (kg)	75.7 ± 7.8	74.1 ± 7.7	69.3 ± 8.7
Fat (%)	13.4 ± 4.7	12.8 ± 4.7	8.2 ± 2.1 $^{<S\ (p = 0.041);\ <R\ (p = 0.016)}$
VO_{2max} (mL/kg/min)	49.9 ± 3.1	58.6 ± 2.9 $^{>R\ (p < 0.001)}$	69.6 ± 4.7 $^{>S\ (p < 0.001);\ >R\ (p < 0.001)}$
$Watts_{max}$	273.13 ± 2.7	312.5 ± 39.3 $^{>R\ (p < 0.001)}$	360.0 ± 40.0 $^{>S\ (p < 0.001);\ >R\ (p < 0.001)}$

Values are expressed as mean SD; superscripts indicate statistically significant differences and their direction. E = elite, S = sub-elite, R = recreational.

2.3. Data Collection

Participants were weighed and measured by an internationally certified anthropometrist (ISAK level 2). Height (cm) was measured using the SECA-360 measuring rod (SECA©, Spain) with a precision of 1 mm, and bodyweight (kg) was measured using scales of the same brand with a precision of 0.1 kg [27]. The tests were carried out between 10 and 12 in the morning. The participants were asked to follow the feeding protocol used for races 3 h before the laboratory appointment. The tests were supervised by a sports doctor and two Bachelor of Science in Physical Activity specialists in performance assessment. The participants performed the entire measurement protocol with their own shorts, slippers, and clipless pedals.

A MyotonPRO® (Myoton Ltd., Estonia) was used to assess the passive stiffness of the main muscles involved in the pedaling action, RF, VL, biceps femoris (BF), and GL, before and after the incremental exercise test. To ensure correct measurements, the assessment points were drawn on the skin following the indications of Hermens et al. [25] (see Figure 1). After removing the tights, and with the cyclist in a lying position on a stretcher, the device was held perpendicular to the skin surface. It was then pushed (0.58 N for 15 ms) against the skin above the muscle area to reach the required depth (d = 3 mm). After the red light turned green, five short impulses (tap interval was 0.8 s) were produced automatically by the device in order to induce mechanical oscillations in the soft tissues. In order to guarantee the validity of the data obtained, only those evaluations in which the coefficient of variation was lower than 3% were considered. Otherwise, the assessment was repeated.

All measurements were made by the same experienced researcher, and the intra- and inter-rater reliability for this device have been estimated in previous studies. [8,21]. The MyotonPRO® device provides data on the recorded passive muscle stiffness (S, N/m) [28]. The mean values for stiffness were calculated from the responses to the five impulses delivered.

Participants performed an incremental exercise test with gas exchange analysis (CPX/D Med Graphics, St. Paul, MN, USA, EE. UU. Measurement accuracy = 1%) [29] in the laboratory. Cyclists completed the graded exercise tests on their own bikes set up on the Wahoo KICKR Power Trainer (Wahoo Fitness, LLC, Atlanta, Georgia), which allows for power and cadence measurements and has been previously validated [30]. The incremental test was based on the following protocol: 10 min of warm up (5 min 100 W + 5 min 150 W) and increases of 25 W every 3 min [31,32]. The test stopped when a plateau of VO_2 was reached or, when not seen, at voluntary fatigue when at 100% of estimated HR_{max}, a respiratory exchange ratio of $\geq$1.15 and a rate of perceived exertion (RPE) of $\geq$18 [32]. The 6–20-point Borg scale was used [33]. All the participants were familiarized with the RPE scale as it was commonly used by their coaches. The scale was shown to the participant in the last 30 s of each one of the steps of the incremental test.

2.4. Statistical Analyses

Statistical analyses were performed in R version 4.0.1 (R Core Team 2020) using RStudio (RStudio Team 2020). Variables were visually inspected and described as mean (standard deviation) using the package *rstatix*. Stiffness differences were assessed by fitting an independent linear mixed-effects model for each muscle (rectus femoris, biceps femoris, gastrocnemius lateralis, and vastus lateralis) using the packages *lme4* and *lmerTest*. The models included fixed-effects terms for time (pre and post), category (elite, sub-elite, and recreational), and their interaction. Time at pre and the recreational category were the reference category in each factor, respectively. Random slopes were allowed to vary between moments (time) and random intercepts were allowed to vary among participants (id). Main effects were obtained performing an analysis of variance with each model, and post-hoc pairwise comparisons were performed comparing estimated marginal means using the emmeans package. The effect size of main effects was reported using partial eta squared (η^2_P) and interpreted as follows: $\eta^2_P < 0.01$ "small", $\eta^2_P < 0.06$ "medium", $\eta^2_P < 0.14$ "large". Differences in estimated marginal means and their 95% confidence intervals were reported as absolute effect size, and Cohen's d with Hedges correction and their 95% confidence intervals were reported as standardized effect size. Cohen's d was interpreted as follows: $|d| < 0.2$ "negligible", $|d| < 0.5$ "small", $|d| < 0.8$ "medium", otherwise "large". Normality of residuals was assessed using the Shapiro–Wilk test and Q–Q plots, heteroscedasticity was assessed using the Breusch–Pagan test, and model performance was evaluated using Akaike information criterion and R^2. All assumptions and performance functions were assessed using the package performance. Statistical significance was assumed when $p < 0.05$.

3. Results

There was a significant correlation between time and category in the VL with stiffness increases in the sub-elite ($p = 0.001$, Cohen's $d = 0.88$) and elite groups ($p = 0.003$, Cohen's $d = 0.72$), but not in the recreational group ($p = 0.085$) (Figure 2). There were no differences among categories in the RF (F(2, 32) = 0.7, $p = 0.53$), GL (F(2, 32) = 0.9, $p = 0.41$), and BF (F(2, 32) = 1, $p = 0.39$). Additionally, baseline stiffness was comparable between categories in all muscles. Stiffness increased over time in both knee extensors, RF (F(1, 32) = 31.9, $p < 0.001$, $\eta^2_P = 0.5$) and VL (F(1, 32) = 24.4, $p < 0.001$, $\eta^2_P = 0.2$), but no changes were observed in the knee flexors, GL (F(1, 32) = 0.2, $p = 0.63$, $\eta^2_P = 0$) and BF (F(1, 32) = 4.1, $p = 0.052$, $\eta^2_P = 0.1$).

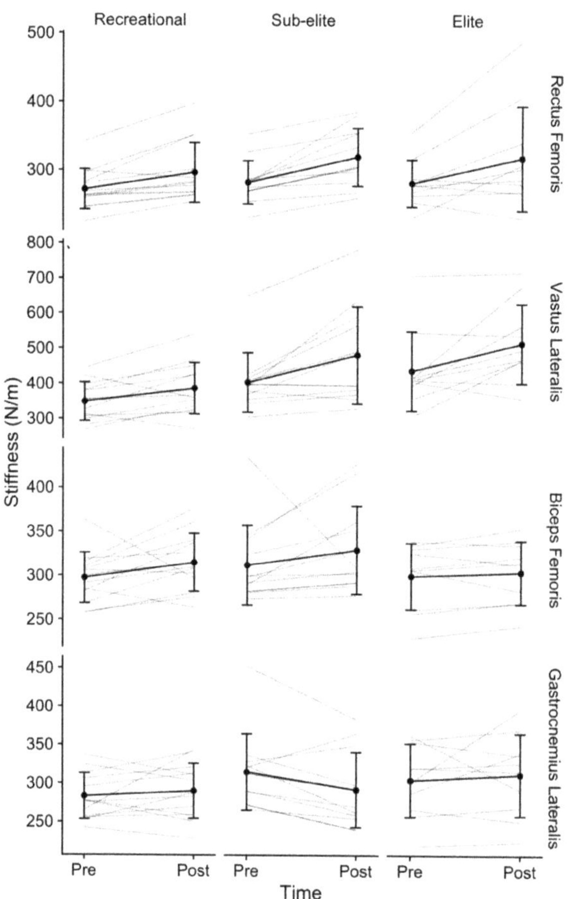

Figure 2. Stiffness by group, muscle, and time. Dots indicate group means whereas vertical error bars indicate standard deviations for each mean.

4. Discussion

The purpose of this study was to analyze the passive muscle stiffness of the main muscles involved in the pedaling action in a group of 35 cyclists classified by performance level. Furthermore, the effect of an incremental exercise test until exhaustion on the variations of passive muscle stiffness was also studied to determine whether performance level has an effect on the post-effort muscular response. The main findings of this study could be highlighted as; (i) there were no differences in the resting passive muscle stiffness of the muscles involved in the pedaling action between cyclists categorized by performance level, (ii) an exposition to an incremental exercise test until exhaustion caused an increase in passive muscle stiffness of the knee extensor muscles regardless of the performance group without resulting in modifications of the knee flexor and ankle extensor muscles, and (iii) only the VL differed in its behavior when differentiating elite and sub-elite categories from recreational cyclists. Therefore, it could be determined that there were no significant differences in the passive muscular stiffness analyzed in a resting situation regardless of the level of the cyclist. Furthermore, the subjection to an incremental test until exhaustion only caused an increase in the stiffness of the knee extensors (RF and VL) with significant differences between performance levels (elite and sub-elite vs. recreational) only found for the VL. To the best of the authors' knowledge, this has been the first study to compare cyclists' muscle stiffness both in a resting and fatigued situation. In addition, the conclusions reported in this manuscript offer information for cyclists, coaches, and medical staff that help to understand the internal behavior of the muscles and could be considered as a training response variable.

The muscular properties and their behavior during the pedaling action have been studied in the scientific literature [34–36]. However, muscle stiffness and especially its variations among cyclists of variable performance levels have been scarcely studied [17,18]. The results obtained in this study differ from those reported by other authors. No significant differences between the analyzed groups (recreational, sub-elite, and elite) or in relation to other parameters such as age, height, or fat percentage in any of the studied muscles (RF, VL, BF, and GL) were obtained. Contrarily, recent studies reported differences in stiffness between performance levels in other sport disciplines. Pruyn et al. analyzed muscle stiffness in netball players classified as elite, sub-elite, and recreational, and reported significant differences between groups ($p = 0.018$). In addition, they concluded by stating that muscular stiffness could be a characteristic that could contribute to a player's ability to physically perform at an elite level. They also provided an explanation for the high injury rates at elite levels of performance by associating greater levels of stiffness with a higher injury rate [37]. Additionally, Kalkhoven et al. established a relationship between greater muscular stiffness and higher performance in a group of soccer players, highlighting the importance of high stiffness and its contribution to better athletic performance [38].

Regarding cycling, the studies performed by Wastford et al. and Uchiyama et al. should be highlighted. Both determined the importance of high stiffness levels for sprint specialties [15,16]. Uchiyama et al. established a mean value of 186–626 N/m in the VL and determined that this value is proportional to the workload and the power developed by the cyclist [16]. Finally, in the only previous study that analyzed cycling stiffness through myometry, higher values were reported for the knee extensor muscles (VL and RF) in sprinters than in less powerful riders. Our results showed that, in a resting situation, the stiffness characteristics are not associated with a typical endurance performance parameter such as VO_{2max}. This finding could be explained because previous studies have analyzed sport disciplines in which the speed component is key to performance [7,37]. This finding does not occur in road cycling, which is a non-impactful discipline characterized by a continuous cyclical movement through coordinated submaximal contractions of the muscles involved in the pedaling action, characteristics that may explain these results. In this study there were no differences in muscle stiffness regardless of the VO_{2max} (348 ± 55 N/m–433 ± 115 N/m). It should be taken into consideration that, despite the lack of statistically significant differences, greater stiffness was observed in those groups with greater aerobic capacity (greater VO_{2max}) in the analysis of the VL (recreational; 348 ± 55, sub-elite; 404 ± 85, and elite; 433 ± 115). This increase was not observed for the rest of the analyzed muscles: RF, BF, and GL.

The literature regarding post-effort stiffness in endurance sports is scarce. Both studies by García-Manso et al. [13] and Andonian et al. [14] on long-distance events (Ironman and ultra-marathon, respectively) determined a clear decrease in contractile capacity and a decrease in the stiffness of the quadricep muscles. The results reported in our work showed an increase in the muscle tone of the knee extensors (RF and VL), with significant differences between both muscle groups ($p < 0.001$). Contrarily to what was reported by previous authors, this increase could be explained because the time until the assessment was clearly different in these studies: 4h in the case of the Ironman and days in the case of the ultra-marathon. In our study, muscle stiffness was tested immediately after the incremental test, which may be considered as a relevant factor that may influence the results.

Our results match those obtained by Silva et al., who determined that the RF and VL were the muscles with the highest activation rates during the pedaling action. Regarding the antagonist muscle (BF), a lower but longer total activation was observed [39]. Two years later, the same authors performed a similar analysis, this time with more muscle groups that were analyzed after an incremental test until exhaustion [40]. Again, activation of both RF and VL increased together with some parts of the hamstring muscles (long head of the BF, semitendinosus, and semimembranosus), while there was no activation in the short portion of the BF. The results of our work determined that the BF did not suffer significant differences in stiffness after undergoing a situation of induced fatigue ($p = 0.052$). This could be due to the fact that most of the power during the pedaling action is produced by the RF and VL and to a lesser degree by the BF. This aspect is closely linked to the elevation of post-effort muscle stiffness [15,41]. Additionally, it should be considered that the technique used for the analysis of activation in these studies was EMG and not myometry, an aspect that could influence the results.

In relation to the GL, only three studies have studied the muscular stiffness of this muscle through myometry [20,35,42]. The participation and activation of this muscle in the pedaling action is indisputable, but our results determined that there were no changes in stiffness in the post-effort situation ($p = 0.63$), an aspect that contradicts the results of some studies that highlight the importance of this muscle group involved in flexion and extension of the ankle. Pruyn et al. studied the relationship between muscle stiffness using myometry and variables related to performance in different modalities of team sports and highlighted the importance of enhancing the muscle group composed by the gastrocnemius, soleus, and Achilles tendon in order to achieve success in these sports modalities [42]. Again, the disciplines analyzed to reach this conclusion were based on short, high-intensity motor actions and not cyclical actions composed of submaximal muscle contractions as occurs in a sport such as cycling.

Finally, the correlation between the moment of measurement (pre and post effort) and the study category should be highlighted. In this case, cyclists in the elite and sub-elite groups presented significant differences in the stiffness values of the VL muscle compared to the recreational group. This finding is related to what was reported by Ditroilo et al., who found that cyclists with higher baseline stiffness levels presented greater reductions in muscle stiffness after fatiguing [18]. This would explain why the lower-level group (lower stiffness and lower VO_{2max}) presented a behavior that did not match what was seen in the higher-level group. This discrepancy between riders characterized by different fitness levels may suggest that the stiffness control of the knee extensor muscles could be useful as a possible reference for functional tests in the periodic evaluation of cyclists.

Despite the fact that the study sample used in this work covered very high levels of performance, the recreational group presented high basal aerobic levels ($VO_{2max} = 49.9 \pm 3.1$ mL/kg/min), and this could be a limitation that had an impact on the results. In the same way, an incremental test generates maximum aerobic metabolic stimulation but does not induce the same levels of structural fatigue. Future lines of research should not only use groups characterized by lower performance levels but also control groups that would allow a global vision of the muscular stiffness. In addition, the behavior of the muscle should be analyzed in more fatiguing situations such as stage races while, at the same time, considering the specialty of the cyclist. Given that the stiffness values of the VL muscle differed between recreational, elite, and sub-elite cyclists, future research should assess whether this marker could be used to differentiate cyclists of varying fitness levels and its potential applicability for the monitoring of training load.

5. Conclusions

The results of this study suggest that there are no differences in the passive muscle stiffness of the muscles involved in the pedaling action between cyclists categorized by performance level. Exposition to an incremental exercise test until exhaustion caused an increase in passive muscle stiffness of the knee extensor muscles, regardless of the performance group, without resulting in modifications of the knee flexor and ankle extensor muscles. Only the VL differed in its behavior when differentiating elite and sub-elite categories from recreational cyclists.

Author Contributions: I.L.-L., R.C.-S., F.C. and S.S. were involved in conceptualizing and design this study. I.L.-L. and S.S. were involved in the data collection. R.C.-S. carried out the statistics. All authors were involved in manuscript writing (review and editing) and supervised this research study. All authors have read and agreed to the published version of the manuscript.

Funding: This research did not receive any external funding.

Institutional Review Board Statement: The study followed the ethical guidelines of the 2013 Declaration of Helsinki and received approval from the Research Ethics Committee of the autonomous region of Aragon, Spain (the approval code is PI19/447, approved on 4 December 2019).

Informed Consent Statement: All participants signed an informed consent document.

Conflicts of Interest: The authors declare no conflict of interest.

References

1. Castronovo, A.M.; Conforto, S.; Schmid, M.; Bibbo, D.; D'Alessio, T. How to assess performance in cycling: The multivariate nature of influencing factors and related indicators. *Front. Physiol.* **2013**, *21*, 116–125. [CrossRef]
2. Lucia, A.; Hoyos, J.; Chicharro, J.L. Physiology of professional road cycling. *Sport Med.* **2001**, *3*, 325–337. [CrossRef]
3. De Pauw, K.; Roelands, B.; Cheung, S.S.; de Geus, B.; Rietjens, G.; Meeusen, R. Guidelines to classify subject groups in sport-science research. *Int. J. Sports Physiol. Perform.* **2013**, *8*, 111–122. [CrossRef]
4. Phillips, K.E.; Hopkins, W.G. Determinants of cycling performance: A review of the dimensions and features regulating performance in elite cycling competitions. *Sport Med.* **2020**, *6*, 23–29. [CrossRef]
5. Lucía, A.; Joyos, H.; Chicharro, J.L. Physiological response to professional road cycling: Climbers vs. time trialists. *Int. J. Sports Med.* **2000**, *21*, 505–512. [CrossRef] [PubMed]
6. Hopker, J.G.; Coleman, D.A.; Gregson, H.C.; Jobson, S.A.; Von der Haar, T.; Wiles, J.; Passfield, L. The influence of training status, age, and muscle fiber type on cycling efficiency and endurance performance. *J. Appl. Physiol.* **2013**, *115*, 723–729. [CrossRef] [PubMed]
7. Gissis, I.; Papadopoulos, C.; Kalapotharakos, V.I.; Sotiropoulos, A.; Komsis, G.; Manolopoulos, E. Strength and speed characteristics of elite, subelite, and recreational young soccer players. *Res. Sports Med.* **2006**, *14*, 205–214. [CrossRef] [PubMed]
8. Zinder, S.M.; Padua, D.A. Reliability, validity, and precision of a handheld myometer for assessing in vivo muscle stiffness. *J. Sport Rehabil.* **2011**, *20*. [CrossRef]
9. Mroczek, D.; Maćkała, K.; Chmura, P.; Superlak, E.; Konefał, M.; Seweryniak, T.; Borzucka, D.; Rektor, Z.; Chmura, J. Effects of plyometrics training on muscle stiffness changes in male volleyball players. *J. Strength Cond. Res.* **2019**, *33*, 910–921. [CrossRef]
10. Dankel, S.J.; Razzano, B.M. The impact of acute and chronic resistance exercise on muscle stiffness: A systematic review and meta-analysis. *J. Ultrasound.* **2020**, *23*, 473–480. [CrossRef]
11. Bojsen-Møller, J.; Magnusson, S.P.; Rasmussen, L.R.; Kjaer, M.; Aagaard, P. Muscle performance during maximal isometric and dynamic contractions is influenced by the stiffness of the tendinous structures. *J. Appl. Physiol.* **2005**, *99*, 986–994. [CrossRef]
12. Watsford, M.L.; Murphy, A.J.; McLachlan, K.A.; Bryant, A.L.; Cameron, M.T.; Crossley, K.M.; Makddisse, M. Prospective study of the relationship between lower body stiffness and hamstring injury in professional Australian rules footballers. *Am. J. Sports Med.* **2010**, *38*, 2058–2064. [CrossRef]
13. García-Manso, J.M.; Rodríguez-Ruiz, D.; Rodríguez-Matoso, D.; de Yves, S.; Sarmiento, S.; Quiroga, M. Assessment of muscle fatigue after an ultra-endurance triathlon using tensiomyography (TMG). *J. Sports Sci.* **2011**, *29*, 619–625. [CrossRef] [PubMed]
14. Andonian, P.; Viallon, M.; Le Goff, C.; de Bourguignon, C.; Tourel, C.; Morel, J.; Giardini, G.; Gergelé, L.; Millet, G.P.; Croisille, P. Shear-wave elastography assessments of quadriceps stiffness changes prior to, during and after prolonged exercise: A longitudinal study during an extreme mountain ultra-marathon. *PLoS ONE* **2016**, *11*, 1–21. [CrossRef]
15. Watsford, M.; Ditroilo, M.; Fernández-Peña, E.; D'Amen, G.; Lucertini, F. Muscle stiffness and rate of torque development during sprint cycling. *Med. Sci. Sports Exerc.* **2010**, *42*, 1324–1332. [CrossRef] [PubMed]
16. Uchiyama, T.; Saito, K.; Shinjo, K. Muscle stiffness estimation using a system identification technique applied to evoked mechanomyogram during cycling exercise. *J. Electromyogr. Kinesiol.* **2015**, *25*, 847–852. [CrossRef]
17. Klich, S.; Ficek, K.; Krymski, I.; Klimek, A.; Kawcznski, A.; Madeleine, P.; Fernández-de-las-Peñas, C. Quadriceps and patellar tendon thickness and stiffness in elite track cyclists: An ultrasonographic and myotonometric evaluation. *Front. Physiol.* **2020**, *11*, 1–12. [CrossRef]
18. Ditroilo, M.; Watsford, M.; Fernández-Peña, E.; D'Amen, G.; Lucertini, F.; de Vito, G. Effects of fatigue on muscle stiffness and intermittent sprinting during cycling. *Med. Sci. Sports Exerc.* **2011**, *43*, 837–845. [CrossRef] [PubMed]
19. Hu, X.; Lei, D.; Li, L.; Leng, Y.; Yu, Q.; Wei, X.; Lo, W.L.A. Quantifying paraspinal muscle tone and stiffness in young adults with chronic low back pain: A reliability study. *Sci. Rep.* **2018**, *8*, 14343. [CrossRef]
20. Huang, J.; Qin, K.; Tang, C.; Zhu, Y.; Klein, C.S.; Zhang, Z.; Li, C. Assessment of passive stiffness of medial and lateral heads of gastrocnemius muscle, Achilles tendon, and plantar fascia at different ankle and knee positions using the MyotonPRO. *Med. Sci. Monit.* **2018**, *24*, 7570–7576. [CrossRef] [PubMed]

21. Bizzini, M.; Mannion, A.F. Reliability of a new, hand-held device for assessing skeletal muscle stiffness. *Clin. Biomech.* **2003**, *18*, 459–461. [CrossRef]
22. Aird, L.; Samuel, D.; Stokes, M. Quadriceps muscle tone, elasticity and stiffness in older males: Reliability and symmetry using the MyotonPRO. *Arch. Gerontol. Geriatr.* **2012**, *55*, 31–39. [CrossRef] [PubMed]
23. Agyapong-Badu, S.; Warner, M.; Samuel, D.; Stokes, M. Measurement of ageing effects on muscle tone and mechanical properties of rectus femoris and biceps brachii in healthy males and females using a novel hand-held myometric device. *Arch. Gerontol. Geriatr.* **2016**, *62*, 59–67. [CrossRef]
24. Klich, S.; Krymski, I.; Kawczynski, A. Viscoelastic properties of lower extremity muscles after elite track cycling sprint events: A case report. *Cent. Eur. J. Sport Sci. Med.* **2020**, *29*, 5–10. [CrossRef]
25. Hermens, H.J.; Freriks, B.; Disselhorst-Klung, C.; Rau, G. Development of recommendations for SEMG sensors and sensor placement procedures. *J. Electromyogr. Kinesiol.* **2000**, *10*, 361–374. [CrossRef]
26. Faul, F.; Erdfelder, E.; Buchner, A.; Lang, A.G. Statistical power analyses using G*Power 3.1: Tests for correlation and regression analyses. Behav. *Res. Methods* **2009**, *41*, 1149–1160. [CrossRef]
27. Witjaksono, F.; Lestari, W.; Sutanto, K.; Wulandari, Y.; Mutu-Manikam, N.R.; Octovia, L.I. Proportion of malnourished ambulatory hospitalized patients based on international criteria at Cipto Mangunkusumo Hospital, Indonesia. *J. Pak. Med. Assoc.* **2021**, *71*, 107–111.
28. Chen, G.; Wu, J.; Chen, G.; Lu, Y.; Ren, W.; Xu, W.; Xu, X.; Wu, Z.; Guan, Y.; Zheng, Y.; et al. Reliability of a portable device for quantifying tone and stiffness of quadriceps femoris and patellar tendon at different knee flexion angles. *PLoS ONE* **2019**, *14*, 1–17. [CrossRef] [PubMed]
29. Gore, C.J.; Catcheside, P.G.; French, S.N.; Bennett, J.M.; Laforgia, J. Automated VO_{2max} calibrator for open-circuit indirect calorimetry systems. *Med. Sci. Sports Exerc.* **1997**, *29*, 1095–1103. [CrossRef]
30. Zadow, E.K.; Kitic, C.M.; Wu, S.S.X.; Smith, S.T.; Fell, J.W. Validity of power settings of the Wahoo KICKR Power Trainer. *Int. J. Sports Physiol. Perform.* **2016**, *11*, 1115–1117. [CrossRef] [PubMed]
31. Mcnaughton, L.R.; Roberts, S.; Bentley, D.J. The relationship among peak power output, lactate threshold, and short-distance cycling performance. *J. Strength Con. Res.* **2006**, *20*, 157–161. [CrossRef]
32. Beltz, N.M.; Gibson, A.L.; Janot, J.M.; Kravitz, L.; Mermier, C.M.; Dalleck, L.C. Graded exercise testing protocols for the determination of VO_{2max}: Historical perspectives, progress, and future considerations. *J. Sports Med.* **2016**. [CrossRef]
33. Borg, G.A. Psychophysical bases of perceived exertion. *Med. Sci. Sports Exerc.* **1982**, *14*, 377–381. [CrossRef] [PubMed]
34. Wakeling, J.M.; Hodson-Tole, E.F. How do the mechanical demands of cycling affect the information content of the EMG? *Med. Sci. Sports Exerc.* **2018**, *50*, 2518–2525. [CrossRef] [PubMed]
35. Wakeling, J.M.; Horn, T. Neuromechanics of muscle synergies during cycling. *J. Neurophysiol.* **2009**, *101*, 843–854. [CrossRef]
36. Sarto, F.; Spörri, J.; Fitze, D.P.; Quinlan, J.I.; Narici, M.V.; Franchi, M.V. Implementing ultrasound imaging for the assessment of muscle and tendon properties in elite sports: Practical aspects, methodological considerations and future directions. *Sports Med.* **2021**, *8*. [CrossRef]
37. Pruyn, E.C.; Watsford, M.L.; Murphy, A.J. Diferences in lower-body stiffness between levels of netball competition. *J. strtrength Cond. Res.* **2015**, *29*, 1197–1202. [CrossRef] [PubMed]
38. Kalkhoven, J.T.; Watsford, M.L. The relationship between mechanical stiffness and athletic performance markers in sub-elite footballers. *J. Sports Sci.* **2018**, *36*, 1022–1029. [CrossRef]
39. da Silva, J.C.L.; Tarassova, O.; Ekblom, M.M.; Andersson, E.; Rönquist, G.; Arndt, A. Quadriceps and hamstring muscle activity during cycling as measured with intramuscular electromyography. *Eur. J. Appl. Physiol.* **2016**, *116*, 1807–1817. [CrossRef]
40. da Silva, J.C.L.; Ekblom, M.M.; Tarassova, O.; Andersson, E.; Rönquist, G.; Grundström, H.; Arndt, A. Effect of increasing workload on knee extensor and flexor muscular activity during cycling as measured with intramuscular electromyography. *PLoS ONE* **2018**, *13*, e0201014. [CrossRef]
41. McLachlan, K.A.; Murphy, A.J.; Watsford, M.L.; Rees, S. The interday reliability of leg and ankle musculotendinous stiffness measures. *J. Appl. Biomech.* **2006**, *22*, 296–304. [CrossRef] [PubMed]
42. Pruyn, E.C.; Watsford, M.; Murphy, A. The relationship between lower-body stiffness and dynamic performance. *Appl. Physiol. Nutr. Metab.* **2014**, *39*, 1144–1150. [CrossRef] [PubMed]

MDPI
St. Alban-Anlage 66
4052 Basel
Switzerland
Tel. +41 61 683 77 34
Fax +41 61 302 89 18
www.mdpi.com

Medicina Editorial Office
E-mail. medicina@mdpi.com
www.mdpi.com/journal/medicina

www.ingramcontent.com/pod-product-compliance
Lightning Source LLC
LaVergne TN
LVHW070617170726
843515LV00004B/105
* 9 7 8 3 0 3 6 5 3 8 6 7 9 *